I0488932

Detection of Nuclear Weapons and Materials: Science, Technologies, Observations

Jonathan Medalia
Specialist in Nuclear Weapons Policy

June 4, 2010

Congressional Research Service

7-5700

www.crs.gov

R40154

CRS Report for Congress
Prepared for Members and Committees of Congress

Summary

Detection of nuclear weapons and special nuclear material (SNM, plutonium, and certain types of uranium) is crucial to thwarting nuclear proliferation and terrorism and to securing weapons and materials worldwide. Congress has funded a portfolio of detection R&D and acquisition programs, and has mandated inspection at foreign ports of all U.S.-bound cargo containers using two types of detection equipment.

Nuclear weapons contain SNM, which produces suspect signatures that can be detected. It emits radiation, notably gamma rays (high-energy photons) and neutrons. SNM is dense, so it produces a bright image on a radiograph (a picture like a medical x-ray) when x-rays or gamma rays are beamed through a container in which it is hidden. Using lead or other shielding to attenuate gamma rays would make that image larger. Nuclear weapons produce detectable signatures, such as radiation or a noticeable image on a radiograph. Other detection techniques are also available.

Nine technologies illustrate the detection portfolio: (1) A new scintillator material to improve detector performance and lower cost. This project was terminated in January 2010. (2) GADRAS, an application using multiple algorithms to determine the materials in a container by analyzing gamma-ray spectra. If materials are the "eyes and ears" of detectors, algorithms are the "brains." (3) A project to simulate large numbers of experiments to improve detection system performance. (4, 5) Two Cargo Advanced Automated Radiography Systems (CAARS) to detect high-density material based on the principle that it becomes less transparent to photons of higher energy, unlike other material. (6) A third CAARS to detect material with high atomic number (Z, number of protons in an atom's nucleus) based on the principle that Z affects how material scatters photons. This project was terminated in March 2009. (7) A system to generate a 3-D image of the contents of a container based on the principle that Z and density strongly affect the degree to which muons (a subatomic particle) scatter. (8) Nuclear resonance fluorescence imaging to identify materials based on the spectrum of gamma rays a nucleus emits when struck by photons of a specific energy. (9) The Photonuclear Inspection and Threat Assessment System to detect SNM up to 1 km away, unlike other systems that operate at very close range. It would beam high-energy photons at distant targets to stimulate fission in SNM, producing characteristic signatures that may be detected. These technologies are selected not because they are necessarily the "best" in their categories, but rather to show a variety of approaches, in differing stages of maturity, performed by different types of organizations, relying on different physical principles, and covering building blocks (materials, algorithms, models) as well as systems, so as to convey many points on the spectrum of detection technology development.

This analysis leads to several observations for Congress. It is difficult to predict the schedule or capabilities of new detection technologies. It is easier and less costly to accelerate a program in R&D than in production. "Concept of operations" is crucial to detection system effectiveness. Congress may wish to address gaps and synergisms in the technology portfolio. Congress need not depend solely on procedures developed by executive agencies to test detection technologies, but may specify tests an agency is to conduct. Ongoing improvement in detection capabilities produces uncertainties for terrorists that will increase over time, adding deterrence beyond that of the capabilities themselves.

This report will be updated occasionally.

Contents

Figures

Appendixes

Contacts

Congress has been interested in detecting nuclear weapons and the special materials needed to make them for many years, especially since 9/11. Nuclear detection has many applications for countering nuclear terrorism and nuclear proliferation, such as securing nuclear weapons and materials in U.S., Russian, and other nuclear facilities, tracking materials at border crossings and choke points, screening maritime cargo containers, and examining actual or suspect nuclear sites.

The United States currently uses several types of nuclear detection equipment. All have significant shortcomings. Some work only at very short range; some cannot identify the material emitting radiation, which can lead to false alarms and interrupt commerce; some depend on operator skill, and may be defeated by a clever smuggler or a sleepy operator; and some are easily defeated by shielding. In an effort to overcome such problems, Congress has funded a pipeline of advanced-technology research, development, and acquisition.

This report seeks to help Congress understand this technology. It discusses the science of detecting nuclear weapons and materials, describes nine advanced U.S. technologies selected to illustrate the range of projects in the pipeline, and offers observations for Congress. The report does not compare technologies.[1] The inclusion of the nine technologies should not be taken to mean that CRS judges them to be better than the hundreds of others not considered here. The report does not discuss the controversial Advanced Spectroscopic Portal because a detailed discussion of it could draw attention from the other technologies considered here.

The scope of this report excludes the organization of the government for dealing with nuclear detection, the role of intelligence and law enforcement in detecting terrorist nuclear weapons, detection of radiological dispersal devices (such as "dirty bombs"), the role of nuclear forensics in attributing an attack to its perpetrator, response to a nuclear attack, and the architecture of a national nuclear detection system.[2] Nor does it discuss possible means by which terrorists might acquire a bomb, or whether they could make a bomb on their own. Much has been written on these topics.[3] While many who are concerned with nuclear detection focus on thwarting nuclear terrorism, this report focuses on technology per se. It avoids extensive discussion of means to counter detection to avoid classified information.

Nuclear detection technology has a dual role in thwarting a terrorist nuclear attack—deterrence and defense. Deterrence means dissuasion from an action by threat of unacceptable consequences. Terrorists may be deterred from a nuclear strike by one of the few consequences unacceptable to them: failure. Detection systems would raise that risk. These systems could also make a terrorist nuclear strike too complex to succeed. But other factors would also have these effects: the difficulty of fabricating a bomb, the chance that law enforcement or intelligence would detect efforts to obtain a bomb, the possible inability to detonate a purloined bomb, and the risk that

---

[1] Comparison would require a detailed review of hundreds of technology projects to determine which are most worthy of further examination; creating metrics to compare the selected projects; and obtaining accurate data for use in the metrics. Each of these tasks would require the work of many experts over many months.

[2] For further information, see CRS Report RL34574, *The Global Nuclear Detection Architecture: Issues for Congress*, by Dana A. Shea.

[3] See Charles Ferguson and William Potter, *The Four Faces of Nuclear Terrorism*, Monterey, CA, Center for Nonproliferation Studies, 2004; Michael Levi, *On Nuclear Terrorism*, Cambridge, MA, Harvard University Press, 2007; and Carson Mark et al., "Can Terrorists Build Nuclear Weapons?," in Paul Leventhal and Yohan Alexander, *Preventing Nuclear Terrorism: The Report and Papers of the International Task Force on Prevention of Nuclear Terrorism*, a Nuclear Control Institute book, Lexington, MA, Lexington Books, 1987, pp. 55-65.

---

scientists recruited for the plot would defect. Such risks would disappear, however, if terrorists were given a bomb and operating instructions. They would then only need to mount a smuggling operation. In that case, the role of nuclear detection systems would change: they would become the main defense.

Chapter 1. Nuclear Weapons and Materials: Signatures and Detection

As background for understanding the detection technologies in Chapter 2, this chapter outlines nuclear detection science. The **Appendix** provides more detail.

What Is to Be Detected?

Detection focuses on weapons and the nuclear materials that fuel them. Weapons can be small. In the Cold War, the United States built high-yield weapons several feet long, atomic demolition munitions that a soldier could carry, and nuclear artillery shells. A terrorist-made weapon would probably be larger.

Nuclear weapons require fissile material, atoms of which can fission (split) when struck by fast or slow neutrons[4]; pieces of this material can support a nuclear chain reaction. The fissile materials used in nuclear weapons are uranium, isotope 235 (U-235), and plutonium, isotope 239 (Pu-239). The Atomic Energy Act of 1954 designates them as "special nuclear material" (SNM).[5] Uranium in nature is 99.3% U-238 and 0.7% U-235; U-235 must be concentrated, or "enriched." Uranium enriched to 20% U-235 is termed highly enriched uranium, or HEU, but nuclear weapons typically use uranium enriched to 90% or so. In this report, HEU refers to this weapons-grade enrichment level. Weapons-grade plutonium, or WGPu, is also a mix of isotopes, at least 93% Pu-239. According to one account by five nuclear weapon scientists from Los Alamos National Laboratory, it would take 26 kg of HEU or 5 kg of WGPu to fuel a bomb,[6] amounts that would fit into cubes 11 cm or 6 cm, respectively, on a side.

Photons 101

Nuclear detection makes extensive use of photons, packets of energy with no rest mass and no electrical charge. Electromagnetic radiation consists of photons, and may be measured as wavelength, frequency, or energy; for consistency, this report uses only energy, expressed in units of electron volts (eV).[7] Levels of energy commonly used in nuclear detection are thousands or

---

[4] Some materials can fission only when struck by fast neutrons; fissile materials are the only materials that can fission when struck by slow as well as fast neutrons.

[5] The Atomic Energy Act of 1954, 42 U.S.C. 2014, defines SNM as uranium enriched in the isotopes 233 or 235 or plutonium. The Nuclear Regulatory Commission has not declared any other material to be SNM even though the Act permits it to do so. U.S. Nuclear Regulatory Commission. "Special Nuclear Material," http://www.nrc.gov/materials/sp-nucmaterials html.

[6] Mark et al., "Can Terrorists Build Nuclear Weapons?"

[7] An electron volt is a very small unit of energy, "a unit of energy equal to the work done by an electron accelerated through a potential difference of 1 volt." http://wordnetweb.princeton.edu/perl/webwn?s=electron%20volt.

millions of electron volts, keV and MeV, respectively. The electromagnetic spectrum ranges from radio waves (some of which have photon energies of 10^{-9} eV), through visible light (a few eV), to higher-energy x-rays (10 keV and up) and gamma rays (mostly 100 keV and up).

X-ray photons and gamma-ray photons of the same energy are identical. However, they are generated in different ways. Gamma rays originate in processes in an atom's nucleus. Each radioactive isotope that emits gamma rays does so in a unique energy spectrum, as in **Figure 1**, which is the only way to identify an isotope outside a laboratory. A detector with a form of "identify" or "spectrum" in its name, such as Advanced Spectroscopic Portal or radioactive isotope identification device, identifies isotopes by their gamma-ray spectra. X-rays originate in interactions with an atom's electrons. Many detection systems use x-ray beams, which can have higher energies than gamma rays and thus are more penetrating. X-ray beams are often generated through the *bremsstrahlung* process, German for "braking radiation," which works as follows. An accelerator creates a magnetic field that accelerates charged particles, such as electrons, which slam into a target of heavy metal. When they slow or change direction as a result of interactions with atoms, they release energy as x-rays whose energy levels are distributed from near zero to the energy of the electron beam. They do not exhibit spectral peaks like gamma rays.

What signatures show the presence of nuclear weapons and SNM?

For purposes of this report, a signature is a property by which a substance (in particular, SNM) may be detected or identified. This section presents several signatures. The **Appendix** and Chapter 2 discuss others.

Atomic number and density

Atomic number, abbreviated "Z," is the number of protons in an atom's nucleus. It is a property of individual atoms. In contrast, density is a bulk property, expressed as mass per unit volume. In general, the densest materials are those of high Z. These properties may be used to detect uranium and plutonium. Uranium is the densest and highest-Z element found in nature (other than in trace quantities); plutonium has a slightly higher Z (94), and its density varies from slightly more to slightly less than that of uranium. Some detection methods discussed in Chapter 2, such as effective Z, make use of Z, and some, such as radiography and muon tomography, make use of Z and density combined.

Opacity to photons

An object's opacity to a photon beam depends on its Z and density, the amount of material in the path of the beam, and the energy of the photons. Gamma rays and x-rays can penetrate more matter than can lower-energy photons, but dense, high-Z material absorbs or scatters them. Thus a way to detect an object, such as a bomb, in a container is to beam in x-rays or gamma rays to create a radiograph (an opacity map) like a medical x-ray.

Radioactivity

Radioactive atoms are unstable and give off various types of radiation; the types of use for nuclear detection are gamma rays and neutrons.

Gamma rays . Gamma-ray spectra are well characterized for each isotope. **Figure 1** and **Figure 2** show spectra for HEU and WGPu. Each point on the spectrum shows the number of photons emitted (vertical axis) at each energy level (horizontal axis). Background gamma radiation is ubiquitous. Since many materials, including SNM, emit gamma radiation, elevated levels of gamma radiation may or may not indicate the presence of SNM, but careful analysis of the total gamma-ray spectrum, as discussed in Chapter 2 under GADRAS, may reveal the presence of SNM. Of particular interest, uranium that has been through a nuclear reactor picks up a very small amount of uranium-232, which decays through intermediate steps to thallium-208. The latter has a half-life of 3 minutes, and its decay produces a gamma ray of 2.614 MeV (as shown in **Figure 1**), one of the highest-energy gamma rays. As a result, it is distinctive as well as highly penetrating, facilitating detection.[8]

Figure 1. Gamma-Ray Spectra: 90% Uranium-235 vs. Background

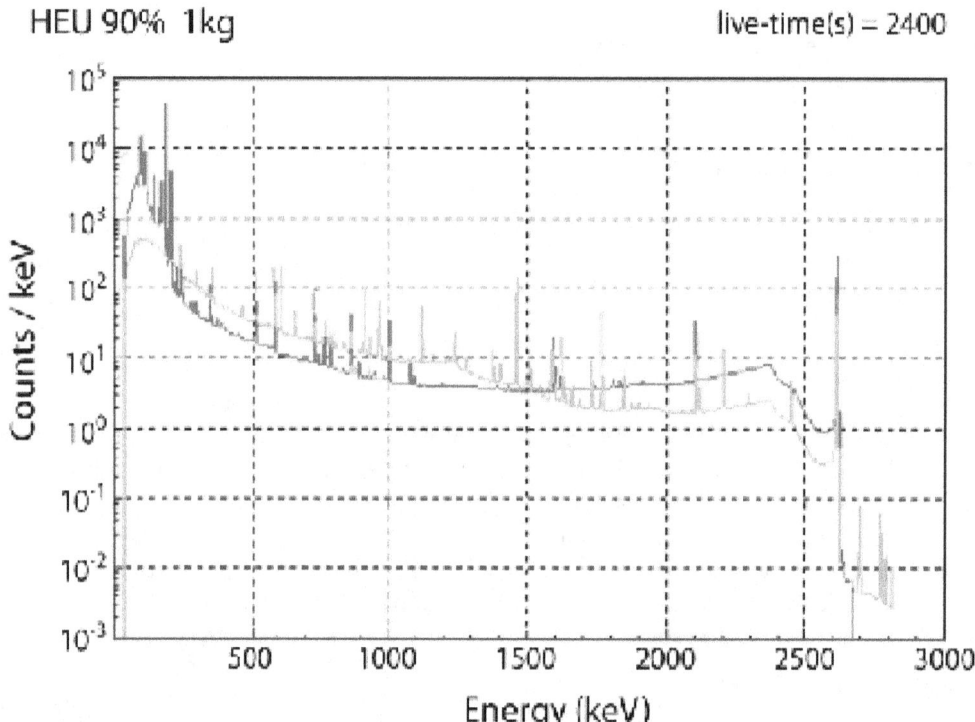

Source: Lawrence Livermore National Laboratory.

Notes: Spectra were taken with a high-purity germanium detector at a distance of 1 m and computed with the GADRAS algorithm, discussed below. Uranium is unshielded, background is dirt. The uranium-235 line starts as the upper line (in red) at far left.

---

[8] For further information on gamma rays generated by uranium, as well as shielding and detection of uranium, see Bernard Phlips et al., "Comparison of Shielded Uranium Passive Gamma-Ray Detection Methods," *Proceedings of SPIE,* vol. 6213, 62130H (2006); doi:10.1117/12.666342, online publication date May 24, 2006; abstract available at http://spiedl.aip.org/getabs/servlet/GetabsServlet?prog=normal&id=PSISDG006213000001621 30H000001&idtype= cvips&gifs=yes&ref=no; and J.R. Lemley et al., "Confirmatory Measurements for Uranium in Nuclear Weapons by High-Resolution Gamma-Ray Spectrometry (HRGS)," Brookhaven National Laboratory, BNL-66293, July 25, 1999, http://www.osti.gov/bridge/product.biblio.jsp?osti_id=750764.

**Figure 2. Gamma-Ray Spectra: Weapons-Grade Plutonium
vs. Background**

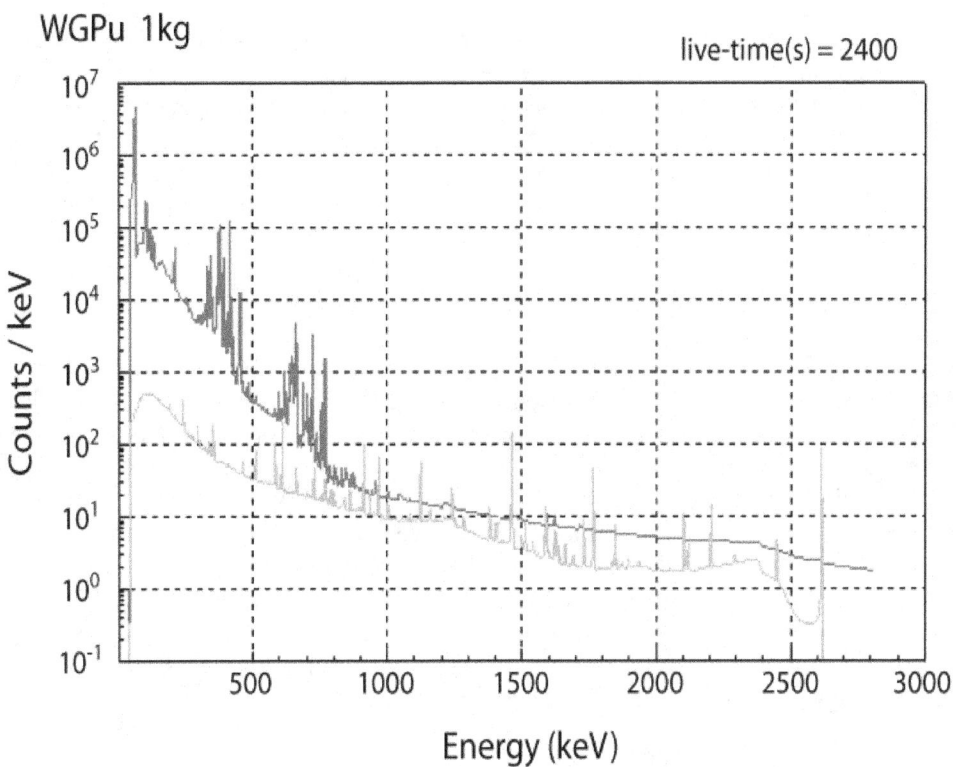

Source: Lawrence Livermore National Laboratory.

Notes: Spectra were taken with a high-purity germanium detector at a distance of 1 m and computer with GADRAS. Plutonium is unshielded, background is dirt. The plutonium line starts as the upper line (in red) at the far left.

Neutrons . Atoms of some heavy elements fission. Of the naturally occurring elements, only U-238 spontaneously fissions at an appreciable rate. Fission releases neutrons. U-235 emits few neutrons, but because U-238, the other main component of HEU, emits some, 1 kg of HEU emits 3 neutrons per second,[9] so it provides a weak signature. Plutonium emits on the order of 60,000 neutrons per kg per second depending on the mix of plutonium isotopes.[10] Unlike gamma rays, neutrons do not have a characteristic energy spectrum by which an isotope can be identified.

How Does Detection Work?

How are signatures gathered, processed, and used?

Detection involves using detector elements to obtain data, converting data to usable information through algorithms, and acting on that information through concept of operations, or CONOPS.

---

[9] Roger Byrd et al., "Nuclear Detection to Prevent or Defeat Clandestine Nuclear Attack," *IEEE Sensors Journal,* August 2005, p. 594.

[10] Ibid.

Detectors, algorithms, and CONOPS are the eyes and ears, brains, and hands of nuclear detection: effective detection requires all three.

Since photons and neutrons have no electrical charge, their energy is converted to electrical pulses that can be measured. This is the task of detectors, discussed next. The pulses are fed to algorithms. An algorithm, such as a computer program, is a finite set of logical steps for solving a problem. For nuclear detection, an algorithm must process data into usable information fast enough to be of use to an operator. It receives data from a detector's hardware, such as pulses representing the time and energy of each photon arriving at the detector. It converts the pulses to a format that a user can understand, such as displaying a gamma ray spectrum or flashing "alarm." Every detector uses one or more algorithms. Improvements to algorithms can contribute as much as hardware improvements to detector capability.

CONOPS may be divided into two parts. One specifies how a detection unit is to be operated to obtain data. Elements include: How many containers must the unit scan per hour? How close would a detector be to a container? Shall the unit screen cargo in a single pass, or shall it be used for primary screening, with suspicious cargo sent for a more detailed secondary screening? A second part details how the data are to be used. Elements include: What happens if the equipment detects a possible threat? Which alarms are to be resolved on-site and which are to be referred to off-site experts? Under what circumstances would a port or border crossing be closed? More generally, how is the flow of data managed, in both directions?[11] What types of intelligence information do inspectors receive, and how do data from detection systems flow to federal, state, and local officials for analysis or action? While this report does not focus on CONOPS because it is not a technology, it is an essential part of nuclear detection.

Principles of detection

Detectors require a signal-to-noise ratio high enough to permit detection. That is, they must extract the true signal (such as a gamma-ray spectrum) from noise (such as spurious signals caused by background radiation). Two concepts are central to gamma-ray detector sensitivity: detection efficiency and spectral resolution. Efficiency refers to the amount of signal a detector records. Radiation intensity (e.g., number of photons per unit of area) diminishes with distance. Since a lump of SNM emits radiation in all directions, using a detector that is larger, or that is closer to the SNM, increases the fraction of radiation from the source that impinges on the detector and thereby increases efficiency. Another aspect is the fraction of the radiation striking the detector that creates a detectable signal. A more efficient detector collects data faster, reducing the time to screen a cargo container.

Spectral resolution refers to the sharpness of peaks in a gamma-ray spectrum. A perfect detector would record a spectrum as vertical "needles" because each radioactive isotope releases gamma rays only at specific energies. Since detectors are not perfect, each energy peak is recorded as a bell curve. The narrower the curve, the more useful the data. Polyvinyl toluene (PVT), a plastic used in radiation detectors that can be fielded in large sheets at low cost, is efficient but has poor resolution. It can detect radiation, but peaks from gamma rays of different energies blur together, which can make it impossible to identify an isotope. **Figure 3** shows the spectra of 90% U-235 and background radiation as recorded by a PVT detector. In contrast, high-purity germanium

---

[11]For further analysis of this topic, see CRS Report RL34070, *Fusion Centers: Issues and Options for Congress*, by John Rollins.

(HPGe) produces sharp peaks, permitting clear identification of specific isotopes. These detectors are expensive, heavy, have a small detector area, and must be cooled to extremely low temperatures with liquid nitrogen or a mechanical system, making them less than ideal for use in the field. **Figure 4** shows the spectrum of Pu-239 as recorded by detectors with better resolution than PVT.

Figure 3. Gamma-Ray Spectra: 90% Uranium-235 vs. Background, Taken with a PVT Detector

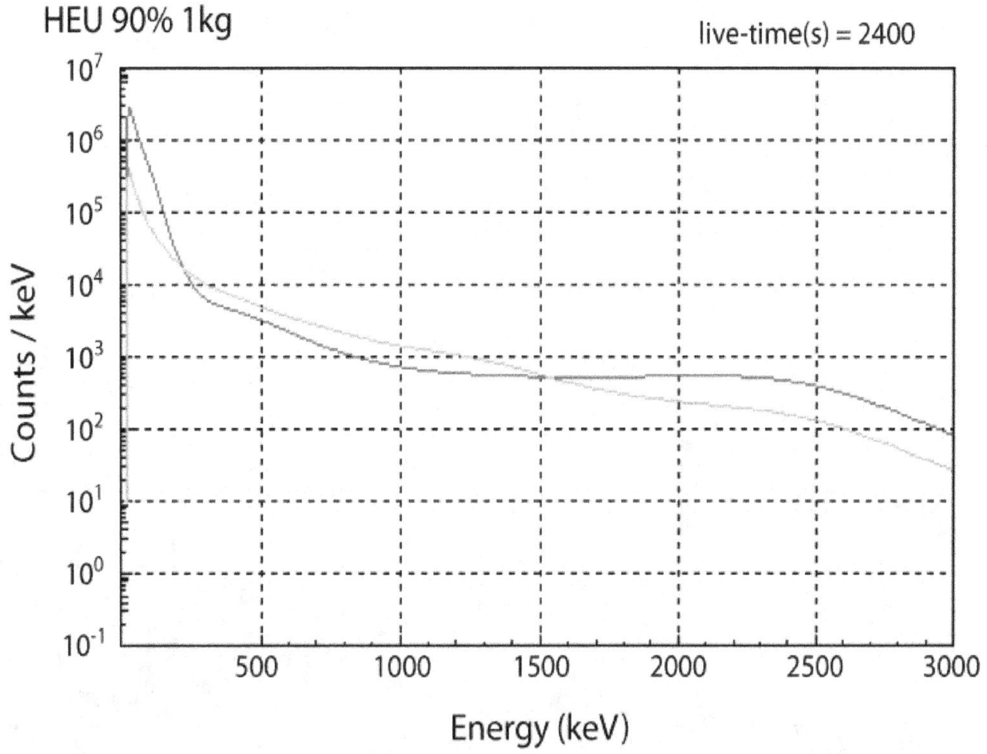

Source: Lawrence Livermore National Laboratory.

Notes: Spectra were taken with a PVT detector at a distance of 1 m and computed with GADRAS. Uranium is unshielded, background is dirt. The uranium-235 line starts as the upper line (in red) at far left.

Figure 4. Gamma-Ray Spectra of Plutonium-239

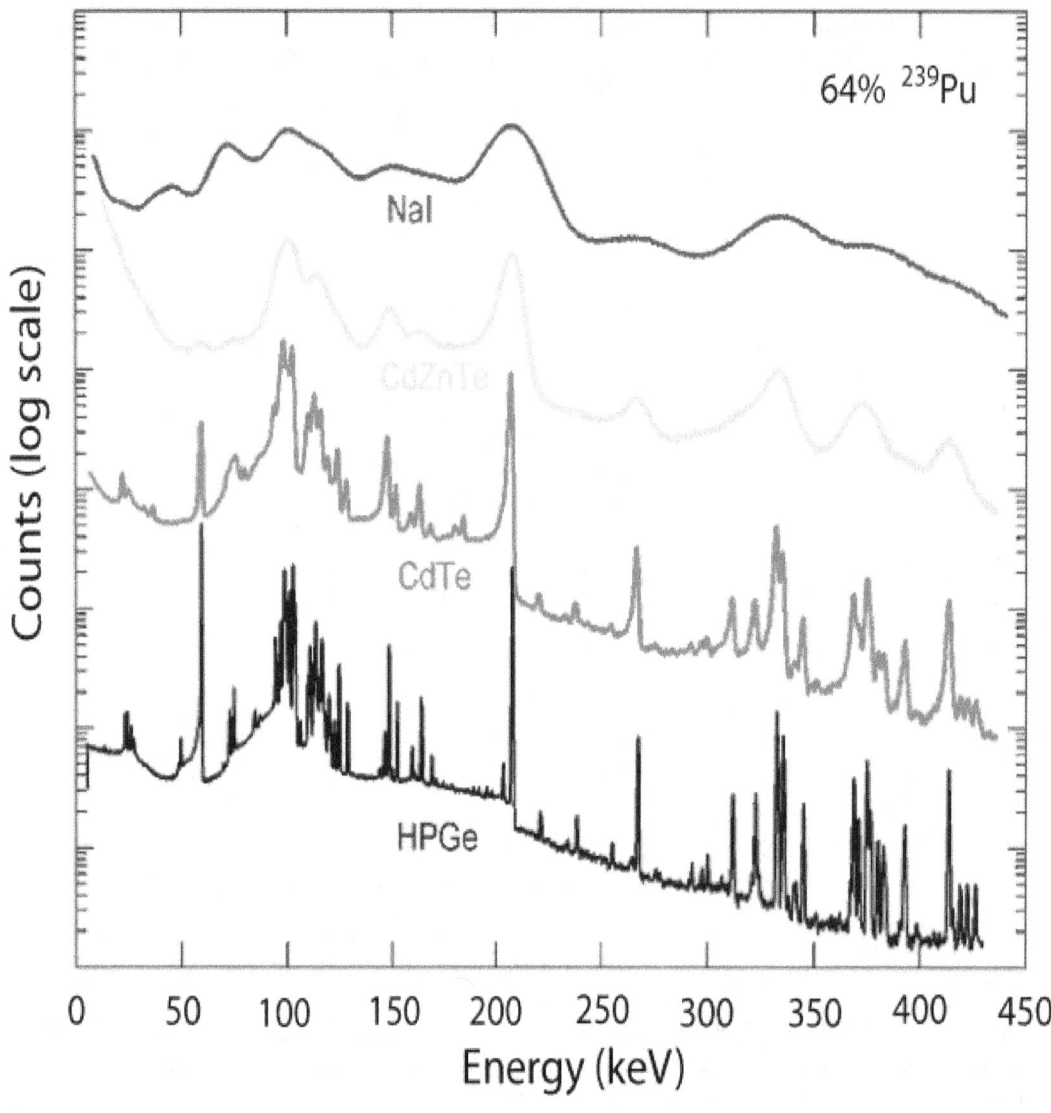

Source: Los Alamos National Laboratory.

Notes: These are spectra of 64% plutonium-239. They were taken using different detector materials to show differences in resolution: from top to bottom, sodium iodide mixed ("doped") with thallium; cadmium-zinc-telluride; cadmium telluride; and high-purity germanium. This figure was created some years ago; the sensitivity of detector materials has improved since then.

Sensitivity can be improved. (1) One type of detector is cadmium-zinc-telluride (CZT) crystals. Better crystals and better ways to overcome their limitations have improved sensitivity. The peak on the right of each spectrum in **Figure 5** shows the cesium-137 spectrum taken with CZT detectors that, for the years indicated, were at the high end of sensitivity. (2) Detectors build radiography images or gamma-ray spectra over time. With more time, a detector can collect more data, in the form of gamma rays or neutrons. More data improve separation of signal from noise and reduce false alarms. Longer scan times improve accuracy but impede the flow of commerce, costing money, so a balance is sought between these two opposed goals. (3) Increasing the spatial resolution of a detector improves sensitivity:

This is easily demonstrated in the example of a shielded versus unshielded radiation detector. Unshielded detectors are sensitive to radiation impinging on it in all directions, which is characteristic of the nature of naturally-occurring background radiation. By adding shielding, a detector's field-of-view can be controlled, and background radiation levels reduced, increasing the signal-to-noise ratio for the detector in the direction the detector is aimed.[12]

---

[12]

Personal communication, Defense Threat Reduction Agency, August 8, 2008.

Figure 5. Resolution of the Cesium-137 Gamma-Ray Spectrum by Different CZT Detectors Has Improved Over Time

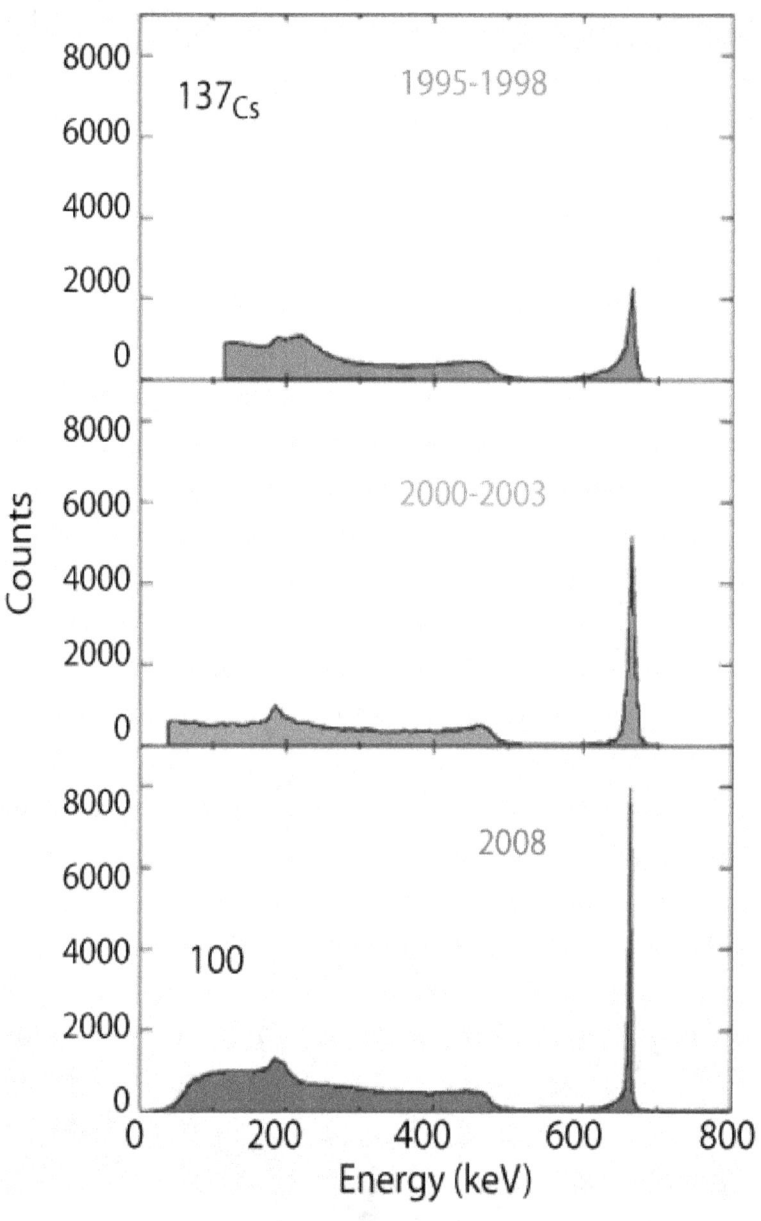

Source: Brookhaven National Laboratory.

Notes: Spectra were taken with cadmium-zinc-telluride (CZT)-based detectors that were high-end in each period listed. The peak on the far right of each spectrum is of particular value for identification of radioisotopes; it becomes sharpers (narrower and higher) with each successive detector. As discussed in the **Appendix**, many factors were responsible for this improvement.

Means of detection

Nuclear detection uses neutrons and photons in various ways. Because either neutrons or photons can readily penetrate most materials, they are the main forms of radiation used to detect radioactive material passively, such as by sensing radiation coming out of a cargo container. Gamma rays and x-rays can be used in an active mode to probe a container for dense material through radiography, which creates an x-ray-type image. Neutrons of any energy level, and photons above about 5.6 MeV, can be beamed into a container to induce fission in SNM. Fission results in the emission of neutrons and gamma rays, which can be detected.

Detecting gamma rays

Gamma rays do not have an electrical charge, but an electrical signal is needed to measure them. There are two main ways to turn a gamma ray into electrical energy. One is with a scintillator material, such as PVT. When a gamma ray interacts with this material, it emits many lower-energy photons of visible light. A photomultiplier tube converts them to electrons, then multiplies the electrons to generate a measurable pulse of electricity whose voltage is proportional to the number of lower-energy photons, which in turn is proportional to the energy deposited by the gamma ray. An electronic device called a multi-channel analyzer sorts the pulse into a "bin" depending on its energy and increases the number of counts in that bin by one. A software package draws a histogram with energy on the horizontal axis and counts on the vertical axis. The histogram is the gamma-ray spectrum for that isotope. In contrast, a semiconductor material, such as HPGe, turns gamma rays directly into an electrical signal proportional to the gamma-ray energy deposited. A voltage is applied across the material, with one side of the material the positive electrode and the other the negative electrode. When a gamma ray interacts with the material, it knocks electrons loose from the semiconductor's crystal lattice. The voltage sweeps them to the positive electrode. Their motion produces an electric current whose voltage is proportional to the energy of each gamma ray. Each pulse of current is then sorted into a bin depending on its voltage and the spectrum is computed as described above.

Detecting neutrons

A common neutron detector is a tube of helium-3 gas linked to a power supply, with positively and negatively charged plates or wires in the tube. In its rest state, current cannot pass through the helium because it acts as an insulator. When a low-energy neutron passes through the tube, a helium-3 atom absorbs it, producing energetic charged particles that lose their energy by knocking electrons off other helium-3 atoms. Positively charged particles move to the negative plate; electrons move to the positive plate. Since electric current is the movement of charged particles, these particles generate a tiny electric current that is counted. Neutron count is an important way to identify SNM because SNM and U-238 emit neutrons spontaneously in significant numbers. Few other sources do. Neutron spectra are of little value for identifying isotopes. They do not have lines representing discrete energies, and neutrons lose energy as they collide with low-Z material, blurring their spectra. However, helium-3 has become extremely scarce, and neutron detection systems for homeland security would require so much of it that alternatives are being sought, such as boron-10.[13]

---

[13] The American Association for the Advancement of Science held a workshop on the helium-3 shortage on April 6, 2010. Briefing slides are available at http://cstsp.aaas.org/agenda_meeting.html.

Detecting dense material

Photons of high enough energy can penetrate solid objects, but are scattered or absorbed by dense objects (or a sufficient thickness of less-dense material). This is the basis for radiography. For cargo scanning, x-rays or gamma rays are beamed through a container, and a detector on the other side records the number of photons received in each pixel. An algorithm then creates an opacity map of the contents. While a bomb would present a sizable image, dense objects in a container might mask a piece of SNM.

Evasion of detection technologies

An enemy could use various means in an effort to defeat detection systems. One such means is shielding. Gamma rays may travel many feet through such low-Z material as wood, food, and plastic, but high-Z material absorbs and deflects them. Conversely, low-Z material absorbs and scatters neutrons, but they pass more readily through higher-Z material. Different amounts of material are needed to attenuate gamma rays depending on their quantity and energy level. Gamma rays from WGPu are sufficiently energetic and plentiful that they are difficult to shield, while a layer of lead would shield gamma rays from HEU, especially if it had not been through a reactor and, in consequence, had not picked up uranium-232, as discussed above. This distinction is of practical significance. Press reports indicate that Iran is using centrifuges to enrich uranium from chemical forms derived from uranium ore, which have not been through a reactor.[14]

Sources of radiation other than SNM complicate detection. Background radiation from naturally occurring radioactive material, such as thorium and uranium, is present everywhere, often in trace amounts. Cosmic rays generate low levels of neutrons. Some commercial goods contain radioactive material, such as ceramics (which may contain uranium) and kitty litter (which may contain thorium and uranium). Other radioactive isotopes are widely used in medicine and industry.

Enemy attempts to defeat one type of detection system may complicate a plot or make it more detectable. (1) It is harder to defeat systems that detect multiple phenomena than a system detecting one phenomenon only. For example, shielding a bomb with lead to attenuate gamma rays would create a large, opaque image on a radiograph. For this reason, Congress mandated that U.S.-bound containers loaded in foreign ports be "scanned by nonintrusive imaging equipment and radiation detection equipment at a foreign port before it was loaded on a vessel."[15] (2) An enemy could attempt salvage fuzing, which would detonate a weapon that sensed attempts to detect it, such as with photon beams, or to open it. However, salvage fuzing could detonate a weapon by accident; if it were scanned overseas; or in a U.S. port, where it would do much less damage than in a city center. (3) Attempts to smuggle HEU into the United States to avoid detection of a complete bomb would require fabricating the weapon inside this nation, which could require such activities as smuggling in other weapon components and purchasing specialized equipment, and could run the risk of accidents, any of which could provide clues to law enforcement personnel.

---

[14] See, for example, Joby Warrick, "Iran's New Centrifuge Raises Concerns about Nuclear Aims," *Washington Post*, May 2, 2010, p. 11.

[15] P.L. 110-53, Implementing Recommendations of the 9/11 Commission Act of 2007, Section 1701, 121 Stat. 489.

Current Detection Technologies

The United States deploys various technologies, such as the following, to detect nuclear weapons or SNM. They are available but have important drawbacks.

Radiation "pagers"

These devices, about the size and shape of a pager, can detect radiation at close distance to alert individuals to the presence of elevated levels of radiation. They may use any of several types of detector material. They are lightweight and inexpensive, but cannot identify the material causing an alarm.

Radiation portal monitors

Many of these devices use large sheets of plastic scintillator material, such as PVT, to detect radiation coming from a vehicle. RPMs were deployed soon after 9/11 because they were available at moderate cost. However, PVT cannot identify the source of radiation. Yet many items in everyday commerce contain radioactive material. As a result, some RPMs produce many false alarms, which may require considerable effort to resolve, delaying the flow of commerce. Newer versions have some isotope identification capability.

Radioactive isotope identification devices

These devices are typically hand-held. They have software that can identify a radioisotope by its gamma-ray spectrum. The most capable of these devices use a crystal of high-purity germanium, a semiconductor material, and are considered the "gold standard" of all identification devices. Such devices are heavy and delicate, and must be cooled with liquid nitrogen or by mechanical means, limiting their usability in the field. They have a relatively short range for detecting radiation sources with low radioactivity, notably shielded HEU, making them unsuitable as the primary method of screening cargo containers.

Radiographic imaging systems

These devices send high-energy photons through cargo containers to create a radiographic image of the contents. The radiograph is scanned, either automatically or by an operator, to search for nuclear weapons, contraband, stowaways, and other illicit cargo. While a nuclear weapon would show up as a white (or black) image on the radiograph and would be clearly visible if hidden in a shipment of low-Z material like food or paper, an operator might overlook it if it were in a shipment of other large, dense objects or jumbled items of various sizes and densities. A small piece of SNM might also be overlooked.

Chapter 2. Advanced Technologies

Many nuclear detection technology projects are under way in the United States and elsewhere. This section discusses nine U.S. technologies selected to include different (1) *agencies sponsoring projects* (the Defense Threat Reduction Agency (DTRA), an agency of the Department of Defense (DOD); the Domestic Nuclear Detection Office (DNDO), an agency of the Department of Homeland Security (DHS); and the National Nuclear Security Administration (NNSA), an agency of the Department of Energy (DOE)), (2) *organizations performing the work* (national laboratories, industry, universities, and collaborations between them), (3) *types of technology* (materials, algorithms, simulation, systems), (4) *physical principles* (muon tomography, radiography, stimulated emission of radiation, nuclear resonance fluorescence), (5) *distances* between the detector and the object being scanned (near and far), and (6) *levels of maturity* (in use for many years but constantly updated, near deployment, and anticipated to be available for deployment in several years). This section does not consider technologies in the earliest stages of development because it is too soon to tell how they will pan out. The discussion of each technology includes several categories:

- The problem that the technology addresses
- Technology background
- Description of the technology
- Potential benefits that the technology offers
- Status, schedule, and funding
- Scientific, engineering, cost and schedule, and operational risks
- Potential gains by increased funding
- Potential synergisms

The last three categories require some explanation.

Risks: The discussion presents potential benefits of the technologies. It does not present "cons." That would be premature because development programs address problems. Instead, each section discusses risk. There are several categories of risk. Scientific problems may thwart a technology. Even if it is scientifically sound, it may be hard to engineer into a workable system. Even if it can be engineered, it may be unaffordable, or take too long to develop. Even if it can surmount these hurdles, problems encountered in field use may render it unattractive.

Potential gains by increased funding: In preliminary discussions, project managers asserted they would, if given more funds, use the added funds to solve problems or exploit opportunities. CRS therefore asked managers of all nine technologies considered in this report how they would spend more money as a way to probe for problems and opportunities with their projects. Would they pursue several promising routes to a technology instead of pursuing only one at the outset? Would they buy more equipment so they could speed up the work? Would they hire more staff?

This analysis applies only to the nine technologies discussed in this report. It is not intended to focus authorization or appropriation consideration solely on them. Other technologies not considered here might realize larger (or smaller or no) gains through increased funding.

Potential synergisms: As the technology descriptions show, many systems have common elements, such as certain types of detectors or algorithms, and work on one technology may contribute to others or have applications beyond current plans.

This report now discusses each of the nine technologies using the above format.

Nanocomposite Scintillators[16]

As of January 2010, DTRA and DNDO terminated the nanocomposite scintillator project. This section, therefore, will not be updated further, but continues in this report because it contains valuable information on nuclear detection. Two small parts of the project, both described below, are continuing as separate projects: the development of application-specific integrated circuits for detecting "thermal" (very low energy) neutrons and identifying gamma-ray spectra simultaneously, and an effort to use nanocomposite scintillator material for neutron detection.

The problem

Scintillator materials are used to detect, and in some cases identify, gamma rays. Higher-performance scintillators are more expensive, harder to manufacture, and fragile; scintillators that are less costly, easier to manufacture, and more rugged offer lower performance. At issue: can the desirable qualities of each type be combined to achieve better performance at lower cost?

Background

Scintillators are materials that, when struck by photons of higher energy, such as gamma rays, capture this energy and release it as photons of lower energy, usually visible light. The material should capture as much of the energy of each photon striking it as possible in order to build an accurate photon energy spectrum and thus identify the material emitting the photons. Ideally, a photon should deposit its full energy in the scintillator material, a so-called full energy interaction. It is also important that the deposited energy be efficiently converted into photons of visible light, which are then counted to determine energy.

There are two main types of scintillators. Inorganic scintillators (those not containing carbon) are typically single crystals, such as sodium iodide (NaI) with a small amount of thallium added. The probability of full energy interaction increases sharply with atomic number (Z) of the scintillator material,[17] and is high for inorganic crystals. The more energy from each photon a scintillator absorbs and then gives off, the better the correlation between energy input and output, and the more precise the spectrum that can be constructed. As a result, a device using an inorganic crystal has a good ability to identify the radioactive material producing a gamma-ray spectrum. There are several drawbacks. The area of a detector that is sensitive to gamma rays is small (limited to the size of a crystal), so the detector must be close to the object to be searched or must scan for longer time so it can receive more gamma rays. They are fragile; dropping one can destroy it. Many inorganic crystals absorb water and are sensitive to light, so they must be protected from environmental conditions. NaI crystals are easy to grow, but cost about $3 per cubic centimeter (cc) of crystal. Other, higher-resolution scintillators are harder to grow, are more costly, and are sensitive to moisture. For NaI, light output varies strongly with temperature, so the temperature must be stabilized or the data corrected.

---

[16] The principal investigator for this project, Edward McKigney, Senior Project Leader, Safeguards Science and Technology Group, Nuclear Nonproliferation Division, Los Alamos National Laboratory, provided detailed information for this section, personal communications, April-August 2008. Others have commented as well to provide alternative perspectives.

[17] Specifically, the number of gamma rays depositing all their energy increases as Z to the 4.5[th] power.

Organic scintillators have the opposite set of properties. They can be made of plastic, such as PVT. As such, they are easy and cheap to make, and are much less fragile than crystals. They can be produced in bulk, making them suitable for deployment in large sheets, such as for radiation portal monitors. On the other hand, since they are composed mostly of hydrogen and carbon, both very low Z elements, they are very inefficient at absorbing the total energy of gamma rays. As a result, as **Figure 3** shows, peaks in PVT-produced gamma ray spectra are indistinct at best, making such spectra of little or no value for identifying radioisotopes.

Technology description

A research project under way at Los Alamos National Laboratory, the Nanocomposite Scintillator Project, seeks to combine the advantages of both types of scintillator materials to overcome the disadvantages of each. The principle is that "nanocrystals," crystals 2 to 5 nanometers in diameter (1 nanometer = 1 billionth of a meter), of certain inorganic scintillator materials can capture most of the energy from photons, thus offering nearly the performance of single large crystals, if packed densely enough in plastic. The resulting mixture also has the desirable features of plastic. The crystals are lanthanum bromide mixed with cerium, or cerium bromide. The modified polystyrene plastic is a scintillator material, so it increases the amount of energy converted to a detectable signal.

In operation, when a gamma-ray photon strikes this material, its energy is absorbed by nanocrystals and the plastic, raising some atoms to a higher energy level. These atoms give off this energy as photons in the visible and near-visible regions of the electromagnetic spectrum ("optical photons"). A photodetector, an electronic component that generates many electrons for each photon it receives, amplifies the signal and converts it from an optical signal to an electronic signal that can be counted. The number of optical photons generated corresponds to the energy level of the photon striking the material. A multi-channel analyzer counts the optical photons, determines the energy level of the photon striking the material, and increases the count of photons of that energy level by one, by this process creating a gamma-ray spectrum.

Edward McKigney, the principal investigator, believes that nanocomposite scintillator material will be able to discriminate between neutrons and gamma rays. He asserts that simulations support this case. The project has obtained data from experiments using the plastic without nanocrystals and nanocrystals without the plastic, and from the literature, and has used these data in simulations. Basic physics calculations also support this case. Neutrons generate protons when they strike the plastic, and gamma rays generate electrons. The plastic responds differently to protons and to electrons; the same is true of the nanocrystals. However, as of August 2008 the project had not conducted experiments demonstrating the ability of the material to detect neutrons and to differentiate between them and gamma rays.

Potential advantages

McKigney states that because the crystals are tiny, growing them is not difficult and is much less costly than growing large whole crystals of these materials. The material can be fabricated at an industrial scale, he says, further reducing cost. He estimates that this material potentially offers the performance of $300/cc material (e.g., lanthanum bromide) at a cost of 50 cents/cc. Because the material acts as a plastic, it is rugged and flexible, and can be made in large sheets, according to McKigney, increasing the sensitivity of a detector using it. It would operate at ambient temperatures.

Status, schedule, and funding

The project has several components. The largest is to develop nanocomposite scintillator material. There are several smaller components: process scale-up; electronics development; detector design; and basic research for a more advanced material. The budget for the entire project is as follows. Early research started in 2004. Los Alamos provided $65,000 in initial funds in FY2005. The laboratory and DNDO provided $1.2 million for FY2006 and $1.6 million for FY2007. Los Alamos, DNDO, and DTRA provided $4.6 million for FY2008 and are projected to provide $5.5 million for FY2009. The project seeks to deliver a small cylinder (1 inch in diameter and 1 inch high) of the material, and to characterize its performance, by December 2008. Anticipated costs are $4 million for FY2009 for the nanocomposite scintillator component. The goal for the end of FY2009 is to have a pilot-scale demonstration of the scintillator material and to start transferring its technology to industry. This material would be optimized for gamma-ray detection. As of July 2009, the schedule for this demonstration had slipped by 6 to 12 months.[18]

Risks and concerns

Scientific risks and concerns

Fabrication of this material requires finding chemicals that can coat the surface of the nanocrystals so they can disperse properly in the plastic while optimizing other properties.[19] The nanocrystals must be packed densely in the plastic to increase sensitivity and resolution. As of May 2008, researchers had reached a packing level of 6% (by volume), with a goal of 50%. It remains to be seen if they can meet this goal. Packing crystals densely in plastic will change some properties of the plastic,[20] so care must be taken to minimize this problem. Another source of risk is that, as of August 2008, the project had not reached high enough packing levels to allow for measurements to determine the sharpness of gamma-ray peaks in spectra generated by this material. The project plans to conduct experiments on this point by December 2008. Such measurements are expected to reduce scientific risk. However, unexpected nanoscale physics could impair energy resolution. In that case, the particle structure would have to be engineered to mitigate the effects; at worst, these effects might degrade performance.

Engineering risks and concerns

(1) When developing the plastic-crystal composite, attention must be paid to ensuring that the material can be made with industrial processes used to manufacture other plastics. (2) The chemical reaction that occurs in manufacturing plastic gives off heat, potentially creating hot spots that would impair the performance of the material. This is not a problem for very small quantities. At the other end of the scale, for industrial production, the problem is well understood

---

[18] Information provided by Edward McKigney, Los Alamos National Laboratory, e-mail, July 21, 2009.

[19] These properties include probability of a full energy deposition interaction; efficiency of converting deposited energy to light, the signal that is measured; transparency to photons generated in this way so the detector volume responds uniformly; and physical robustness.

[20] For example, dense packing of nanocrystals may change the material's mechanical flow. That would make it harder to process the material by the lowest-cost method, extrusion, but it could still be cast and machined to shape. PVT scintillator is cast and can be machined, suggesting that that approach offers low cost while producing large sheets of material.

---

and chemical engineering solutions have been implemented for decades. A concern is whether a solution can be devised for pilot-scale production.

Cost and schedule risks and concerns

The project is not mature enough to provide a firm estimate of the cost of the material produced on an industrial scale. The cost estimate cited above is based on the cost of procuring the raw materials and assumes that the energy cost for processing is low. Inflation in energy would not be expected to increase cost sharply, but cost of the product is sensitive to the cost of cerium, a rare earth element. The manufacturing processes are similar to those used for such consumer goods as fabric softeners and disposable plastic water bottles, so unit cost arguably should not be high. However, the cost estimate excludes the cost of fixed infrastructure; how much that would add to unit cost depends on infrastructure cost and the number of units produced. Transferring the technology to a commercial partner by the end of FY2009 depends on resolving potential scientific and engineering problems in a timely manner and finding the right partner. The project is not far enough along to make a firm estimate of schedule beyond FY2009.

Operational risks and concerns

In any project, it is possible to develop a product only to have it fail in everyday use. This project is trying to minimize this risk. It is trying to design robustness into the material, such as by (1) using a plastic that is soft and rubbery rather than brittle, (2) ensuring that the material will be effective across the temperature range to which it will be exposed, and (3) ensuring that the detector and packaging are compatible so that, for example, the detector material will not expand so much as to crack its casing.

Potential gains by increased funding

The project's total budget is about $5.5 million per year. McKigney states that he could "usefully employ a total budget of up to $12M/year to reduce the time to deployment of these technologies" in several ways. (1) The project is pursuing, with about one-third of its total funding, a separate basic research project to develop a scintillator material approaching the resolution of high-purity germanium detectors with the cost and processing characteristics of plastic. More funds could accelerate this project, according to McKigney. He cautions that this project might take a decade or more and has much greater scientific risk than the current nanocomposite scintillator project. (2) The project uses equipment available at Los Alamos, but he states that staff could save time if they had their own equipment.[21] (3) It would be difficult to fabricate a single large panel (e.g., 50 cm wide by 20 cm thick by 90 cm long) of detector material. Further, a panel segmented into tens to thousands of tiny panels produces the optimum tradeoff between cost and effectiveness and can provide data on the position of a radioactive source. Each panel element would need its own electronics channel, so application-specific integrated circuits must be custom-designed, which could take two or three years. With more funding, he asserts, the project could develop these

---

[21] For example, the project needs an understanding of the structural features of the materials being synthesized that an instrument called a resonant Raman spectrometer could provide. McKigney states that this instrument would speed up development of the process to synthesize the scintillator material by providing direct information about what material has been synthesized. Currently, the project infers such information from measurements that McKigney states are harder to interpret.

chips and the scintillator material concurrently, saving time. Recognizing this leverage, DTRA provided about $150,000 for this purpose spread over 2½ years beginning in May 2008. McKigney states that additional funding would accelerate this schedule. (4) Hiring more chemists, electrical engineers, and others would accelerate these projects, according to McKigney.

Potential synergisms and related applications

(1) This material offers the greatest benefit in detectors that use large panels of scintillator material, such as some under development as discussed below. (2) By offering greater sensitivity and greater resolution, this material could provide better data for algorithms to process, permitting the development of simpler, more capable algorithms. (3) Nanocomposite scintillator material may be able to measure neutron and gamma ray energies. As such, it might be possible to use it in a detection system instead of separate means of detecting each particle type. (4) Some companies are considering systems that use tubes filled with helium-3 gas to detect neutrons. However, helium-3 is scarce, and there may not be enough to support large-scale use of these tubes. Nanocomposite scintillator material might be an alternative for neutron detection.

GADRAS: A Gamma-Ray Spectrum Analysis Application Using Multiple Algorithms[22]

The problem

In a carefully controlled laboratory environment, a radioisotope can be readily identified by matching its gamma ray spectrum to one in a library of spectra. At a port or border crossing, a spectrum would be complicated by radiation from many sources and by attenuation caused by cargo and other materials. At issue: How can the signal from SNM be extracted from a gamma ray spectrum, and how can this capability be improved?

Background

Nuclear detection hardware receives much attention, but data from the hardware—e.g., pulses of various energies from gamma rays—are unintelligible until processed by software. Indeed, software in the form of algorithms is the "brains" of a detector. An algorithm, such as a computer program, is a finite set of logical steps for solving a problem. For nuclear detection, an algorithm must process data into usable information fast enough to be of use to an operator.

Algorithms are designed to assure a low rate of false positives, which impede commerce, and a near-zero rate of false negatives, which could let a terrorist weapon into the United States. For gamma-ray identification, an algorithm creates equations to model "radiation transport," the movement of radiation through material. A detector will record gamma rays from all sources— background radiation, items in ordinary commerce, radioisotopes that terrorists might include in a shipment to confuse analysis, and a nuclear weapon or SNM. Many uncertainties affect an

---

[22] The principal developer of this algorithm, Dean Mitchell, Distinguished Member of the Technical Staff, Contraband Detection Organization, Sandia National Laboratories, provided detailed information for this section, personal communications, April-August 2008. Others have commented as well to provide alternative perspectives.

algorithm. Gamma rays lose energy as they interact with matter, altering their spectra depending on the type and thickness of the materials they pass through, and the initial energy of a gamma ray. Random fluctuations in the number of counts in each gamma-ray energy bin create statistical uncertainties, especially if the number of counts is low. The composition and thickness of materials between the radiation source and the detector (cargo, shielding, container wall, air, etc.) is not known. The equations may not exactly represent the detector dimensions or the shielding configuration because approximations are made to reduce complex environments to a set of equations that can be more readily computed.

Technology description

Dean Mitchell of Sandia National Laboratories has developed an applications package, "Gamma Detector Response and Analysis Software," or GADRAS.[23] Development of GADRAS started in 1985 for use in the Remote Atmospheric Monitoring Project (RAMP), which used low-resolution detectors to analyze airborne radionuclides.[24] Beginning in 1996, automated isotope identification was developed within GADRAS to process spectroscopic data collected at border crossings to support interdiction of radioactive materials.[25] Work on the current version of GADRAS started in 2001. In earlier versions, it was seen as acceptable to spend a month analyzing an individual spectrum; in the wake of 9/11, in order to be of value for screening commerce, GADRAS had to be modified to process data quickly while minimizing false positives and false negatives. As of April 2010, the GADRAS application included six radiation analysis algorithms, gamma ray and neutron detector response functions, and support for radiation transport calculations.[26]

To identify radioactive sources creating a gamma-ray spectrum, GADRAS matches an entire gamma ray spectrum against one or more known spectra. **Figure 6** illustrates this approach. Many other algorithms focus on peaks in gamma ray spectra because they are the most obvious features. In contrast, GADRAS analyzes the full spectrum for several reasons. (1) Peaks may overlap, making source identification ambiguous. (2) Most counts in a gamma-ray spectrum are often outside the peaks, in which case using only peak data would ignore most of the data. For example, less than 3% of the counts in a spectrum for U-238 occur in the 1.001-MeV peak, the most prominent feature of its spectrum. (3) Counts outside the peaks help characterize the composition and thickness of intervening material. Since gamma rays interact with these materials, characterizing the materials improves the accuracy with which the gamma-ray spectrum as read by a detector can be linked back to the gamma-ray source. Arriving at a solution consistent with all the data increases confidence in the result. (4) The absence of counts in a region of a spectrum can be a clue to the identity of radioactive materials. (5) Using the entire spectrum helps analyze data from scintillators having low energy resolution because low resolution often precludes identification of peaks in the spectrum, and helps analyze spectra of weak sources.

---

[23] For a technical discussion of GADRAS, see Dean J. Mitchell, "Variance Estimation for Analysis of Radiation Measurements," Sandia Report SAND2008-2302, April 2008.

[24] Dean J. Mitchell, "Analysis of Chernobyl Fallout Measured with a RAMP Detector," SAND87-0743-UC-32, Sandia National Laboratories, 1987.

[25] Dean J. Mitchell, "Analysis of Low-Resolution Gamma-Ray Spectra by Using the Unscattered Flux Estimate to Search an Isotope Database," Systems Research Report, Sandia National Laboratories, 1997.

[26] Information provided by Dean Mitchell, e-mail, July 7, 2009, and April 1, 2010.

Figure 6. How GADRAS Identifies Special Nuclear Material

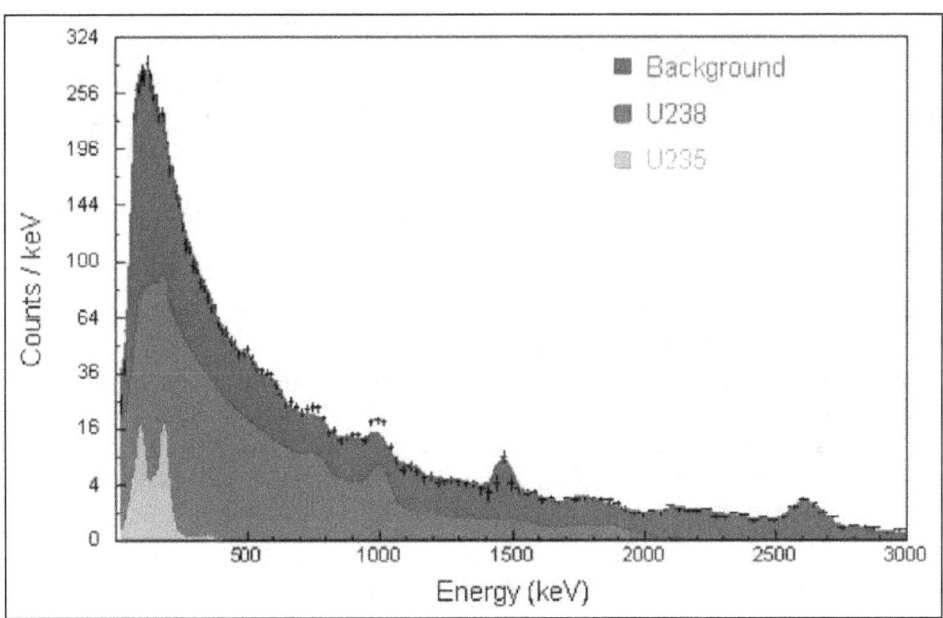

Source: Figure by Dean Mitchell, Sandia National Laboratories, April 2010.

Notes: This figure shows how GADRAS can identify various materials (in this case uranium) despite background radiation. The bottom scale shows gamma-ray energy in thousands of electron volts (keV); the left scale shows number of gamma-ray counts at each keV level. Black bars at the upper edge of the blue area are the raw data that a gamma-ray detector provides. (Figures in the online version of this report are in color.) Data are presented as bars rather than points to indicate one standard deviation uncertainties. The raw data cannot differentiate between gamma rays produced by different materials, as a 200-keV gamma ray from one substance is identical to a 200-keV gamma ray from another substance, and background radiation may hide gamma-ray peaks from a material of interest. A count of gamma rays from a cargo container might produce a data set like this.

GADRAS analyzes the entire spectrum to determine which combination of materials might produce it. In this example, GADRAS finds uranium-238 (red area, middle band) and uranium-235 (green area at bottom left) as well as background radiation (blue area, top band). The key point is that analysis of the entire spectrum provides more accurate identification of the materials under investigation than would analyzing only gamma-ray peaks because much the information in the spectrum is outside the peaks. For example, uranium-238 has a peak at 1,001 keV but in this figure only a very small part of the uranium-238 gamma-ray spectrum (less than 3 percent), and an even smaller fraction of the total, is at that energy level.

GADRAS also uses neutron flux data. Since neutrons pass more readily through high-Z material and gamma rays pass more readily through low-Z material, different materials, such as in a container, will affect the total radiation output differently. As a result, using both gamma ray and neutron data improves the analysis.

GADRAS has been in use since 1986. Since 9/11, more operators have used it in a wider range of applications. In response, the software is continually upgraded to support new types of radiation sensors, provide new capabilities, and meet new performance requirements. According to Mitchell, "One of the new features is the ability to support the analysis of data that is collected with neutron multiplicity counters. This capability enables inclusion all of the data collected by gamma-ray detectors and various types of neutron detectors into a unified analysis algorithm."[27] One goal is to make GADRAS more automated so that less skill is required to operate it. Another goal is to make it faster. A current effort focuses on improving sensitivity to SNM and reducing the false alarm rate. In the past, some sacrifices were made to fidelity of the analysis in order to gain speed; now, with faster computers, it should be possible to improve the analysis while increasing speed; this approach is being investigated. Another approach to reducing false alarms is to examine radiation sensor data collected in 2005 on cars entering the Lincoln Tunnel between New York City and New Jersey. According to Mitchell, the data included 50 to 60 false alarms. Mitchell and other researchers at Sandia are trying to determine what caused the false alarms in order to modify GADRAS to correct for these problems.[28]

The main application of GADRAS is to support "Triage/Reachback" analysis. A radiation detection operator in the field, such as a Customs and Border Protection (CBP) officer, who finds a vehicle or cargo container that presents a suspicious radiation signature that cannot be easily resolved, can send the detection data (such as a gamma ray spectrum) to the Laboratories and Scientific Services section of CBP for a more detailed analysis.[29] That analysis uses GADRAS. Similarly, if that service is unable to resolve the matter, it can send the data to a secondary Reachback at the nuclear weapons laboratories, which also use GADRAS.

Potential advantages

By analyzing a complete gamma ray spectrum, GADRAS increases the accuracy of determining whether a cargo container or other item is carrying SNM, reducing the risk of false positives and false negatives. Using neutron count data in addition to gamma ray spectra further improves capability.

Status, schedule, and funding

GADRAS has been used for cargo inspection since 1998. Software upgrades are released every two months or so. There is no line item for GADRAS development. Mitchell estimates that Sandia is spending perhaps $600,000 in FY2010 to continue to develop GADRAS. DTRA stated in 2008, "The DTRA in concert with NNSA is currently proposing development of the next generation of GADRAS as part of a potential [memorandum of understanding] between the organizations. The effort would emphasize the revision algorithms, update to Vista compliant software, and increasing portability for field use."[30]

---

[27] Information provided by Dean Mitchell, e-mail, April 1, 2010. A neutron multiplicity counter detects SNM by detecting the time pattern of neutron generation. A subcritical mass of highly enriched uranium or weapons-grade plutonium can support a fission chain reaction producing an increasing number of neutrons. (Since it cannot support enough fissions to create a nuclear explosion, such chain reactions die out.) These neutrons are generated in a closely spaced pattern over a brief time. In contrast, background neutrons occur in a random time pattern.

[28] Information provided by Dean Mitchell, e-mail, April 1, 2010. A neutron multiplicity counter detects SNM by detecting the time pattern of neutron generation. A subcritical mass of highly enriched uranium or weapons-grade plutonium can support a fission chain reaction producing an increasing number of neutrons, though it cannot support enough fissions to create a nuclear explosion, so such chain reactions die out. These neutrons are generated in a closely spaced pattern over a brief time. In contrast, background neutrons occur in a random time pattern.

[29] For further information, see U.S. Department of Homeland Security. Customs and Border Protection. "Laboratories and Scientific Services." Available at http://www.cbp.gov/xp/cgov/trade/automated/labs_scientific_svcs/.

[30] Personal communication, August 5, 2008.

Risks and concerns

Scientific risks and concerns

Since upgrades are ongoing, developers face such scientific problems as how to improve equations to characterize the response of detector material to radiation. Such problems are not a major risk to continued development of GADRAS, Mitchell states. Another risk is that the gamma-ray signal from shielded HEU may be so low that even a high-quality algorithm cannot identify the HEU.

Engineering risks and concerns

The major risk to GADRAS development comes from the programming language that is used for the graphical user interface (GUI). The GUI for the current version of GADRAS is written in a Microsoft language called Visual Basic Version 6 (VB6). VB6 functions under current versions of Windows, including the most recent, Vista. Microsoft, however, no longer maintains VB6, so it is not necessarily compatible with new compilers,[31] often leaving no effective upgrade path for large application programs like GADRAS. User feedback indicates that some GADRAS components do not function properly in Vista. While minor changes may resolve the latter problem, the current GUI may not run under a future version of Windows.[32] GADRAS is a large program, and current funding does not support the effort that would be required to update it.

Cost and schedule risks and concerns

Since upgrading GADRAS is a low-budget activity, it entails little cost risk. No near-term requirements impose significant schedule risks. However, it would take one to two years to revise the GUI. Making this revision in advance of a change of operating systems would enable users to use GADRAS without interruption when they switch to the new operating system.

Operational risks and concerns

Since GADRAS has been in existence for many years, and since it is used mainly by scientists and technicians rather than by operators in the field, there is little risk that it will fail in everyday use. The risk that an upgrade will cause a problem is reduced by careful testing. Other risks are the possibility that GADRAS would not run on future versions of Windows operating systems, and that conversion would not be made before an operating system that would not run GADRAS is introduced. Some have said that GADRAS is a complicated program to use; accordingly, the CONOPS for GADRAS is that it is used mainly for Reachback rather than in the field.

---

[31] A compiler is a computer program that translates an application (such as GADRAS) into instructions that a computer can process.

[32] As another example, because of incompatibilities between VB6 and current generations of Fortran, components of GADRAS that are written in Fortran must be compiled using Compaq Fortran, another discontinued and unsupported programming language.

Potential gains by increased funding

Mitchell states that perhaps $600,000 a year for two years would allow Sandia to hire a few programmers who could convert GADRAS to use current compilers in order to avoid potential disruptions associated with new operating systems. A related task is documentation of the Application Programs Interface. This documentation would enable other users, such as at other laboratories, to develop new applications that can access capabilities that are incorporated into the Dynamic Link Library, which performs most of the computations in GADRAS. This increased access would reduce the cost and development time for new applications.

Potential synergisms and related applications

(1) GADRAS might improve the performance of systems that induce fission and detect the resulting radiation. (2) By gaining more information from the gamma ray spectrum, it might reduce the gamma ray or neutron flux needed to induce fission and the dose resulting from fission, thereby increasing worker safety. (3) More computation power would permit more sophisticated iterations of GADRAS to be developed, or would permit GADRAS to run faster, or both. (4) Increased computation power, especially smaller and more capable computers, might enable GADRAS to be modified for use in the field with radiation detection equipment, permitting quicker resolution of suspicious containers and vehicles. Reachback would then be used only for the hardest-to-resolve cases. (5) GADRAS could be modified to improve the performance of radioisotope identification devices (RIIDs). DTRA states that it "is currently funding development of a more portable version of GADRAS ... that is intended to be resident on certain RIID systems, such as handheld HPGe system produced by Ortec."[33] (6) Improved scintillator materials can be expected to provide better data inputs to GADRAS, enabling a further reduction in false negatives and false positives. Universities, companies, and government laboratories are working to develop such materials.

Computer Modeling to Evaluate Detection Capability[34]

The problem

Developing equipment to detect terrorist nuclear weapons and SNM requires many choices. It would be of great value to evaluate how they affect cost and performance before committing to a system. However, many combinations and tradeoffs are possible, and it would be prohibitively expensive and time-consuming to run enough trials for each to make a valid comparison among them. When faced with similar choices, such as in designing a car, corporations typically run huge numbers of simulated trials using computer models that take significant investment to develop and maintain. At issue: How can computer modeling contribute to the development of nuclear detectors, and what are its limitations?

---

[33]Personal communication, August 5, 2008.

[34] Richard Wheeler, Lead for Homeland Security Analysis, Global Security Directorate, Lawrence Livermore National Laboratory, provided detailed information for this section, personal communications, April-August 2008. Others have commented as well to provide alternative perspectives.

Background

A detector should maximize the probability of detecting an actual threat (a true positive) while minimizing the probability of detecting a nonexistent threat (a false alarm, or false positive). For a given technology, these objectives cannot be achieved simultaneously—an improvement in either one comes only at the expense of the other. A receiver operator characteristic (ROC) curve, such as **Figure 7**, illustrates this relationship.[35] By relating the true positive and false positive rates, the curve defines the performance of a receiver (in signal processing, where the term "ROC" originated) or of nuclear detection equipment. ROC curves show that the probability of a true positive and a false positive go up together. This is logical; one could eliminate false alarms by turning off the detector, or could be sure of detecting every threat by having the detector alarm whenever the slightest trace of radiation is detected, which, given omnipresent background radiation, would be all the time. The peril of failing to detect an actual threat is clear. At the same time, law enforcement and commercial interests are intolerant of false alarms because these alarms require a major effort, diverting officers to the scene and possibly unloading a cargo container or closing a border crossing. Further, in the real world, numerous false alarms may cause operators to ignore all alarms or set the detector to be less sensitive, reducing the false alarm rate but also increasing the likelihood of missing an actual threat. This tradeoff is shown in **Figure 7** by moving from point C, with a high probability of detection but a high false alarm rate, to point B, with intermediate values for both, to point A, with a low false alarm rate but a low probability of detection.

Figure 7.A Notional Receiver Operating Characteristic (ROC) Curve

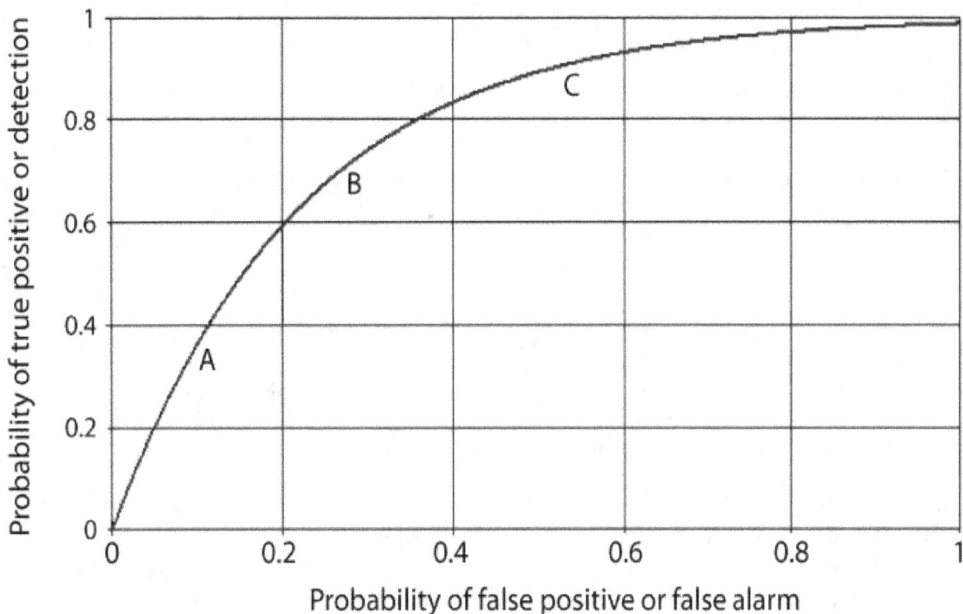

Probability of false positive or false alarm

Source: Prepared by CRS with the assistance of Lawrence Livermore National Laboratory.

---

[35] For a helpful interactive tutorial on ROC curves, see Anaesthetist.com, "The Magnificent ROC," at http://www.anaesthetist.com/mnm/stats/roc/Findex htm.

Figure 8 shows three ROC curves to illustrate several concepts. (1) Moving from curve A to curve B to curve C shows the performance of a hypothetical detector improving over time, perhaps as a different detector material is used or an algorithm is modified. The improvement can be visualized by moving upward (line 1), which shows an increase in the true positive rate for a given false positive rate, or by moving from right to left (line 2), which shows a reduction in the false positive rate for a given true positive rate. (A diagonal line from lower right to upper left would show improvement in both.) (2) A, B, and C could represent differences in performance of one detector under different conditions, such as changes in the background, different operating conditions (e.g., scan time), or different benign materials in a container. (3) The curves could characterize the performance of three competing detectors. Note that actual ROC curves have more complex shapes than the notional curves shown. They may even cross over each other, indicating that one option is not uniformly better than another, requiring consideration of further tradeoffs.

Figure 8. Three Notional ROC Curves

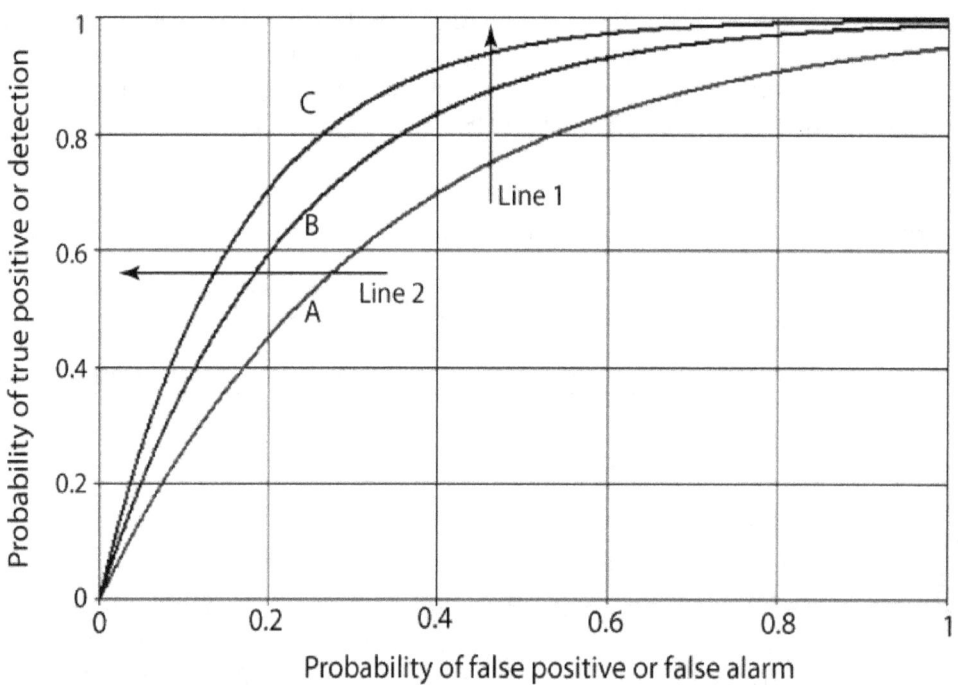

Source: Prepared by CRS with the assistance of Lawrence Livermore National Laboratory.

Many variables affect detector performance. Some are operational, such as scan time, the detector's target (containers, cars, trucks, trains), the distance between detector and target, and environmental conditions (background radiation, season, temperature). The detector is to detect SNM or nuclear weapons, yet the signatures on which it will focus may be accompanied by radiation from innocent sources and background radiation, and may be shielded inadvertently or deliberately. There are choices for the active elements of a detector, the algorithm used, and specific subroutines. These choices affect detector performance. Many combinations of these variables are possible. To gain enough data to make a ROC curve statistically valid, many trials would need to be performed for each combination of controllable variables (operating setup, detector, and algorithm) against many targets, each generating its own radiation signature. Each event in which a vehicle or container passes through the detection system is called an "encounter." It would be prohibitively expensive and time-consuming to run many encounters for each of thousands of combination of controllable variables, but it would be of great value to have the resulting data in order to compare, improve, or choose between detection systems and their components.

Computer modeling can help. Modeling creates mathematical representations of the real world, varies the inputs, and calculates the outputs. In the case of nuclear detection, the real world consists of controllable variables (operational scenario, detector, and algorithm) and uncontrolled variables (signals from radioactive material). The performance of a modeled detector can be illustrated using ROC curves as described above. Running the model to simulate many encounters for each combination of controllable variables provides enough observation points to generate a statistically significant ROC curve. This process is repeated for many combinations of variables. The resulting data show the user how change to variables affects performance, and permit comparing one detector against another. Computer-generated "data" for nuclear detection encounters are always imperfect, as discussed below. As a result, the adequacy with which the models represent reality is always at issue, and model developers devote great effort to improving that representation.

Modeling can also be used to evaluate requirements for elements of a detection system. For example, a model was used to study the spectral resolution (see "Principles of detection," above) required of a detector material to distinguish the gamma-ray spectra of threat sources from non-threat sources. According to a report on this project, "To capture the range of gamma-ray sources and shielding configurations found in commerce, we generated simulated populations of 1000 or more spectra, each with 3000 energy channels."[36] Clearly, it would have been costly and time-consuming to generate the data experimentally.

Technology description

DNDO has established an ongoing program, Detection Modeling and Operational Analysis (DMOA), that the national laboratories and private sector contractors carry out. It builds models that characterize the variables noted above, i.e., the operational scenario, the radiation signals, the detector, and the signal-processing algorithm. Creating mathematical representations of the first three takes an immense amount of work because each is so complex. Alternative algorithms are simulated as part of the overall detection performance modeling. (Algorithm development requires a great deal of work, but is not within the scope of DMOA, which focuses on simulations.) For example, modeling a gamma-ray spectrum requires taking into account various sources of radiation and types of shielding. A DMOA study of real-world data found that the spectra from cargo differed between spring-summer and fall-winter; the study speculated that a different mix of products shipped in the two periods caused the difference.[37]

---

[36] Karl Nelson, Thomas Gosnell, and David Knapp, *The Effect of Gamma-Ray Detector Energy Resolution on the Ability to Identify Radioactive Sources*, Lawrence Livermore National Laboratory, Radiological and Nuclear Countermeasures Program, LLNL-TR-411374, February 2009, p. i, https://e-reports-ext.llnl.gov/pdf/370769.pdf.

[37] Lawrence Livermore National Laboratory, Radiological and Nuclear Countermeasures Program, "Radiation Detection Modeling and Operational Analysis: Benchmarks, Nuisance Source, Algorithm and Resolution Studies," LLNL-TR-401531, December 31, 2007, by Padmini Sokkappa et al., pp. 21-22 This source is a report marked by Lawrence Livermore National Laboratory, the originating organization, "For Official Use Only." That laboratory has approved CRS use of the information cited by this footnote.

The model processes data on the operational scenario, radiation signals, and the detector to create a gamma-ray spectrum that is sent to the algorithm. The algorithm does not "know" the difference between a spectrum generated in the real world or by computer, and processes both in the same way. Since the difference is the source of the data, the key to simulation is generating the gamma-ray spectrum (signal plus noise) that goes to the algorithm. The greater the fidelity with which the model mimics real-world inputs, the better it represents system performance. Modeled performance, as measured by detection probability and false alarm rate, can be summarized in a ROC curve in the same way as empirically measured performance. The modeling effort also includes assessing the sensitivity of detector performance to changes in operational scenarios, detector hardware, and algorithms.

One DMOA effort in 2006 focused on four main areas to improve its models.[38] This effort illustrates the work undertaken to improve the fidelity of models and how this work requires detailed knowledge of the components being modeled.

- Evaluation benchmarks. Providing a basis for comparing systems requires a reference set of threat objects and shielding. Previously, this set included plutonium, HEU, and other threats, as well as different levels of shielding. In 2006, DMOA added new threats and shieldings to the reference set. It also developed a reference set of objects for detection by radiography.

- Background and nuisance source modeling. According to DMOA, "Background radiation and nuisance source population characteristics generally dictate detection threshold settings through their impact on innocent alarm rates. Characterization of these factors is critical to evaluating the performance of radiation detection systems." DMOA used real-world data to develop a model of the distribution of naturally occurring radioactive material. Data from the model were then compared with another set of real-world data to check the validity of the model.

- Operational analysis. DMOA used real-world data to compare the performance of several algorithms. One result was to "highlight the sensitivity of system performance to the detection algorithm used."

- Detector resolution study. DMOA studied the resolution needed to distinguish threats from other sources of radiation. The analysis, though preliminary, found that there is a benefit by having better resolution than that offered by sodium iodide, but that improving resolution beyond a certain point would offer marginal benefit.

Potential advantages

Modeling offers advantages compared to obtaining data from the real world . (1) A model permits construction of statistically significant ROC curves. It is not uncommon to require a false alarm rate of one in ten thousand or lower; statistical validation of performance at this level would require hundreds of thousands or millions of experiments. It would be costly and time-consuming to conduct field trials using a single combination of variables in order to build one statistically significant ROC curve, let alone conducting field trials for thousands of combinations of variables

---

[38] This paragraph is based on ibid., pp. i-ii.

that represent the range of conditions. A model permits running many thousands of simulated encounters in less than a day to explore the range of encounters a detector may face. Once the model has been constructed, validated, and implemented, the cost of these computer runs is very low. (2) A model and a test program are complementary. The test program generates data for the model, and the model can steer the test program by indicating what tests would be of greatest value for improving the predictive power of the model. (3) A model permits exploration of encounters that might occur in the real world but that could not be conducted because of cost, safety, or difficulty. For example, it would be unthinkable to place a nuclear weapon in commercial traffic to test detectors, but the only facilities where such testing could be done, notably the Nevada Test Site, have characteristics very different than those of ports or border crossings. As another example, it might be desirable to see whether a protracted period of heat or high humidity would impair how a detector would function, but it would be much easier to replicate such conditions through modeling. (4) A model permits comparison of components of an encounter, such as which algorithm better processes the data for a given gamma-ray spectrum. (5) A model permits analysis of alternatives and clarification of tradeoffs before committing to a specific design, helping to inform billion-dollar decisions.

Status, schedule, and funding

DMOA is an ongoing program. When DNDO was established, DMOA became an explicit element within the System Architecture program. It has been funded within that program at approximately $2 million per year. The amount funded by all DNDO offices on related detection modeling is on the order of five times that of the Architecture program. Other detection modeling is done for other purposes by other offices within DNDO, as well as by DOE and DOD. However, since modeling activities are inherently cross-cutting and support many technology development and assessment projects, it is difficult to estimate total spending on modeling in the federal budget.

Risks and concerns

Scientific risks and concerns

While no model can be perfect, the key risk is that the model's output (for example, gamma ray spectra that the model generates) might differ significantly from data that would have been obtained from actual field trials. Part of this risk concerns modeling the underlying science. Some aspects are known in detail, such as the radiation spectrum of U-235. But there are uncertainties. HEU is not pure U-235, so the spectrum will be somewhat different from that of U-235. Some shielding can be modeled precisely, such as a centimeter of lead. But the defense (the modeling design team) does not know what shielding, if any, the attacker might use. A cargo container can hold many types of cargo, each of which interferes with radiation transport differently. Different arrangements of the cargo place different amounts and types of material between source and detector, affecting gamma ray spectra. Any model makes approximations and simplifications, sometimes to allow the simulations to run faster, or simply because more fidelity for certain phenomena are judged unnecessary or unwarranted. This creates further risk that the model might not sufficiently represent reality. There are other risks. Items included in the model may not be selected accurately. Systematic errors may arise, such as differences between spring-summer and fall-winter cargo. Adding detailed radiation sources with additional shielding, types of cargo, and detector details may increase the realism of the model, but also add complexity, opening the door to additional errors.

Engineering risks and concerns

Models are complex mathematical approximations of reality. Yet real-world data are often limited, perhaps covering too narrow a range of conditions. DMOA uses statistical methods to transform a real-world data set into data for different detectors and conditions. As a DMOA report states, "LLNL [Lawrence Livermore National Laboratory] has developed a procedure for generating a statistical model of a nuisance source population ... based on measured data. ... The statistical model developed provides a basis for simulating an unlimited number of random samples in nuisance sources for a population assumed to be similar to that underlying the measured data."[39] A difficulty is in validating the model. Various isotopes cause background radiation, and DMOA observes that "there is no simple standard procedure to compare multivariate populations."[40]

Models may also manipulate data by adding in, or "injecting," spectra from well-characterized radiation sources like HEU or WGPu to computed spectra of cargo typical of normal commerce to see how well an algorithm can detect threat material in cargo. However, that approach may be an inadequate representation of reality. It is arguably unlikely that terrorists would include a nuclear weapon or SNM in a random cargo container if they chose that vector; rather, they might arrange the cargo to help evade detection. Simulation developers have not modeled such deliberate arrangements of cargo because that would require thinking about how to model adversary behavior, something outside their expertise. It might be desirable to have terrorism experts modify some injections to reflect adversary behavior.

Cost risks and concerns

Most funds spent on DMOA are for staff salaries, so the likelihood of unanticipated cost increases appears low. However, this assumes that computing hardware required to run the models quickly is already paid for.

Schedule risks and concerns

If terrorists planned to bring a nuclear weapon or SNM into the United States, they would presumably try to evade detection, leading to an offense-defense competition. It would be essential for this nation to stay ahead. That task involves much effort by many entities. For modeling, the task would be to upgrade means of evasion included in the model in time to stay ahead of the threat. This would depend in part on inputs from intelligence services, such as data on specific threats, but also on advances in characterizing background radiation, developing algorithms, developing statistical means of transforming or validating data, and the like. The program is paced by resources available. The risk is that progress in nuclear detection, including modeling, will not be fast enough to defeat the threat.

---

[39] Radiological and Nuclear Countermeasures Program, "Radiation Detection Modeling and Operational Analysis," p. 17. This source is a report marked by Lawrence Livermore National Laboratory, the originating organization, "For Official Use Only." That laboratory has approved CRS use of the information cited by this footnote.

[40] Ibid., pp. 20-21.

Operational risks and concerns

Law enforcement and commercial interests are intolerant of false alarms. The false alarm rate often drives detector settings and choice of detection systems, and increasing the true positive rate also increases the false positive (false alarm) rate. While a model can explore the tradeoff and optimum balance between true and false positive rates, a risk is that systematic errors, incorrect data transformations, and incomplete accounting for background radiation would cause the model to generate a setting that is less than optimal. In some cases, this problem can be overcome by making adjustments in the field, but in other cases, such as the choice of detector material, adjustments may not be possible.

Potential gains by increased funding

At present, the DMOA efforts at the national laboratories and private sector contractors are distributed throughout their organizations. Most staff members who work on detection modeling do so only part time. Some modeling tools (simulation codes, databases, and algorithms) tend to be somewhat informal, developed by various groups for specific needs. According to Richard Wheeler, Lead for Homeland Security Analysis, Global Security Directorate, Lawrence Livermore National Laboratory, added funding in this area could enable the designation of full-time staff, development of standardized modeling tools, acquisition of dedicated computing resources, establishment of rigorous peer review of the models, formal coordination of related efforts sponsored by DOE, DOD, and other agencies, and generation of new databases essential for model validation. While analysis of detector performance against realistic threats is sensitive or classified, many advances in modeling and modeling tools could be made in an open environment, including the academic community. Wheeler states that more funds could support a more substantial engagement of university researchers.

Potential synergisms and related applications

Simulation does not detect anything by itself, but is by its nature synergistic with other aspects of nuclear detection. It has as its purposes improving many elements of detection systems, whether deployed or under development; helping to integrate hardware and software into a system; optimizing systems and CONOPS; and informing decisions on the most cost-effective mix of systems to acquire.

L-3 CAARS: A Low-Risk Dual-Energy Radiography System[41]

The problem

Current radiography systems have important limitations in their ability to detect SNM in cargo containers. A small amount of dense material may be inconspicuous if shipped in a container filled with dense or random objects. SNM could also be placed inside dense objects for camouflage. Another difficulty of detecting SNM in containers is that an operator may fail to see

---

[41] Joel Rynes, Program Manager, CAARS Program, Domestic Nuclear Detection Office, Department of Homeland Security, provided detailed information for this section, personal communications, April-August 2008. Others, including L-3 Communications Corporation, have commented as well to provide alternative perspectives.

a threat. At issue: Can radiography take advantage of different characteristics of SNM to detect it in cargo? And can a system evaluate radiographic images automatically to reduce dependence on operator judgment?

In an effort to overcome these limitations, DNDO is conducting R&D on advanced radiography systems under the Cargo Advanced Automated Radiography System (CAARS) program. CAARS involves three technical approaches to radiography, each with a different contractor. (As noted below, DNDO terminated the contract for the American Science and Engineering, Inc., system in March 2009.) This section and the next two discuss these systems. It must be emphasized at the outset that these systems are under development. As a result, no deployable CAARS systems exist, so diagrams and performance specifications presented in these sections must be viewed as goals that are yet to be demonstrated in commercially available equipment. Customs and Border Protection (CBP) raises several concerns about CAARS, as described under "operational risks and concerns," below. CBP, a component of the Department of Homeland Security, is a front-line agency whose officers perform such missions as operating border crossings and inspecting cargo entering the United States at seaports, airports, and land crossings.

Background

Two types of commercially available equipment for scanning cargo containers have been widely used since 2002.[42] One type, radiation portal monitors, passively detects radiation coming from a container. Another type, radiography equipment, creates x-ray-type images of a container; examples include the Rapiscan Classic Eagle and the SAIC VACIS.[43] This type is more relevant to CAARS. Radiography equipment works as follows. A cargo container is driven between two components of the equipment, or the equipment moves over a stationary container. One component produces gamma rays from a highly radioactive substance like cobalt-60 or cesium-137, or x-rays from a linear accelerator. These photons emerge in a thin vertical fan-shaped beam and pass through the side of the container.[44] On the other side of the container, an array of detector material records photon intensity levels, which correlate to opacity, pixel by pixel. A computer assembles the pixels into a radiograph. A black or white spot indicates an area that is opaque to photons of the energy used. Key limitations of such systems are that they cannot

---

[42] "Prior to 9/11, not a single radiation portal monitor [RPM] and only 64 large-scale non-intrusive inspection [i.e., radiography] systems were deployed to our nation's borders. By October of 2002, CBP had deployed the first RPM at the Ambassador Bridge in Detroit." Testimony of Thomas Winkowski, Assistant Commissioner, Office of Field Operations, U.S. Customs and Border Protection, before the Senate Homeland Security and Governmental Affairs Committee, September 25, 2008. CBP has 1,145 RPMs and 209 large-scale radiography systems deployed as of October 23, 2008. Personal communication, Patrick Simmons, Director, Non-Intrusive Inspection, Customs and Border Protection, October 23, 2008. Radiography has been used for industrial applications (e.g., inspecting metal parts for cracks and voids) for decades.

[43] For information on these systems, see Rapiscan Systems, "Rapiscan Eagle P6000," at http://www.rapiscan.com/eagle-P6000.html, and Science Applications International Corporation (SAIC), "Safety & Security: VACIS Cargo, Vehicle, and Contraband Inspection Systems," at http://www.saic.com/products/security/.

[44] Jonathan Katz et al. argue that it would be more effective to have the beam interrogate a cargo container vertically rather than horizontally. "In innocent cargo long slender dense objects are packed with their longest axes horizontal, and dense cargoes are spread on the floor of the container. Therefore, near-vertical irradiation will only rarely show regions of intense absorption [of photons] in innocent cargo. In contrast, horizontal irradiation would often find this 'false positive' result, requiring manual unloading and inspection. Another advantage of downward near-vertical illumination is that the Earth is an effective beam-stop, combined with a thin lead ground plane, its albedo [the fraction of the beam that is reflected] is negligible and additional shielding would not be required." J.I. Katz et al., "X-Radiography of Cargo Containers," *Science and Global Security,* Vol, 5, No. 1, January 2007, pp. 49-56.

---

differentiate between different types of material and cannot pick out threats from clutter. While a complete nuclear weapon might or might not be noticed, detection probability decreases as the threat object becomes smaller, so it would be difficult for current radiography systems to detect a small piece of HEU.[45]

A different physical principle enables a radiography system to search for materials with high atomic number (Z). Both the L-3 and SAIC CAARS systems utilize this principle. They use one or two linear accelerators to generate 6-MeV electrons and x-rays with energies from 0 to 6 MeV and, in the same manner, x-rays with energies from 0 to 9 MeV (abbreviated in this section as 6-MeV x-rays and 9-MeV x-rays). The former have the greatest number of photons at around 2 MeV; the latter, at about 3 MeV. (The third CAARS system utilizes a different principle.)

Photons of these two energies interact with matter differently. Six-MeV x-rays scatter when they strike electrons.[46] The amount of scattering is a function of the electron density of the material, so it increases with Z. As a result, high-Z material is more opaque to 6-MeV x-rays than is low-Z material, and creates a bright or dark spot on a radiograph. **Figure 9**, top panel, is a radiograph taken with 6-MeV x-rays. Nine-MeV x-rays interact more strongly with an atom's nucleus. When a nucleus absorbs a photon, it releases energy in the form of an electron-positron pair. This effect is proportional to Z squared. Thus high-Z material is much more opaque to 9-MeV x-rays (many fewer get through the material being interrogated because they are absorbed) than to 6-MeV x-rays.

---

[45] Katz states, "The real issue in radiographing thick targets is not the source strength, but scattering in the target. If you don't have a narrow Bucky collimator matched to the geometry, the signal (increase of attenuation of unscattered photons showing the high-Z object ...) is swamped by forward-scattered photons from the mass of solid material. The small feature disappears from the image (rather like exposing undeveloped camera film to the light)." Personal communication, August 8, 2008. For cargo screening using radiography, a "Bucky collimator" is a piece of high-Z metal (such as tungsten) placed just in front of a photon detector element. The metal has a small hole precisely aligned with the direction of the interrogation beam in order to eliminate most photons that have been scattered by the cargo and that could obliterate small features of the image, such as a piece of SNM. The resulting increase in signal-to-noise ratio greatly improves the detector's ability to pick out suspicious cargo.

[46] The scattering mechanisms are much more complicated than can be described here. For details, see Glenn Knoll, *Radiation Detection and Measurement,* third edition (New York, John Wiley & Sons, 2000), pp. 48-53, 308-312.

Figure 9. Dual-Energy Radiography

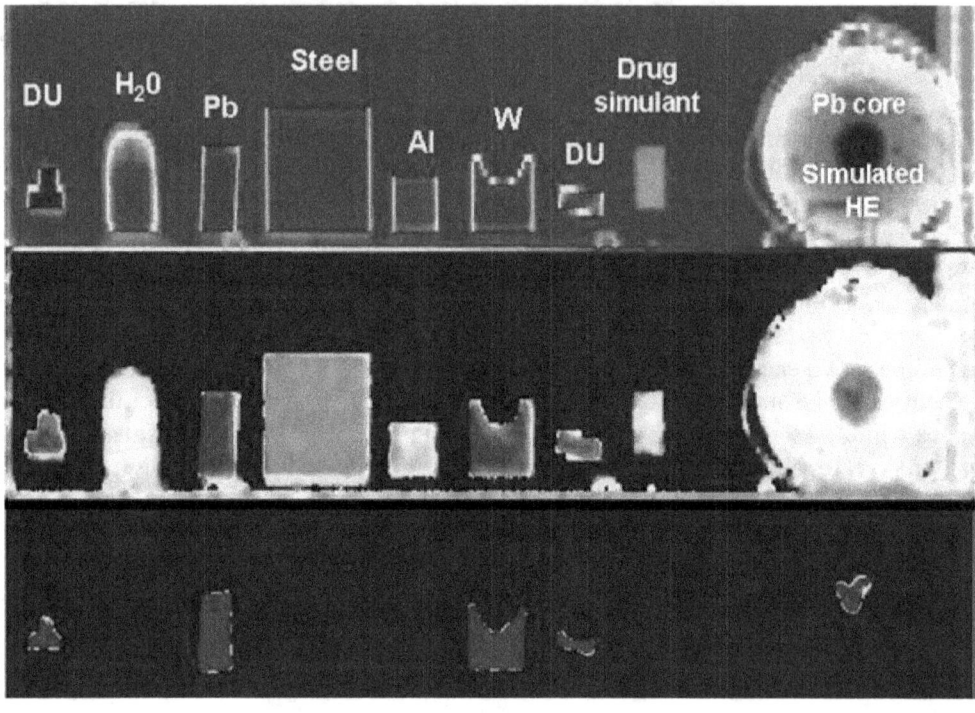

Source: These images were taken in January 2006 with an SAIC proof-of-concept laboratory prototype cargo imaging system called VACIS-Z: Detection of High-Z Material in Cargo. The VACIS-Z contract was funded by the Homeland Security Advanced Research Projects Agency of the Department of Homeland Security through a contract that predates DNDO (which was established in April 2005). The contract number is N41756-04-C-4200, awarded under Technical Support Working Group (TSWG) Broad Agency Announcement DAAD-03-T-0024. For further information on TSWG, see http://www.tswg.gov/.

Notes: This graphic shows three radiographs of the same objects. (Abbreviations: DU, depleted uranium, i.e., natural uranium with most uranium-235 removed, used here as a surrogate for highly enriched uranium; Pb, lead; Al, aluminum; W, tungsten; HE, chemical high explosive) The top image was taken using *bremsstrahlung* photons with a maximum energy of 6 MeV. The middle image shows, pixel by pixel, the ratio of a 9-MeV and a 6-MeV radiograph. Higher-Z objects are more opaque to photons with energies up to 9 MeV than to those with energies up to 6 MeV, and pixels of such objects are represented in darker colors. The bottom image shows only the high-Z material from the middle image, which would automatically generate an alarm. In this image, the system would alarm on depleted uranium, lead, tungsten, and lead surrounded by a simulated chemical high explosive.

The L-3 and SAIC CAARS exploit this difference to create a so-called dual-energy radiograph, in which each pixel represents the *ratio* of opacity of that pixel to 9-MeV and 6-MeV x-rays. To create the radiograph, an algorithm assigns different colors to different ratios; in the middle panel of **Figure 9**, higher-ratio pixels are darker. Pixels with ratios above a certain value indicate high-Z material. While there is no absolute physical threshold between medium- and high-Z material, CAARS uses Z>72 as the threshold for high-Z material;[47] elements with Z>72 include tungsten, gold, lead, uranium, and plutonium.[48] Pixels of such material could be presented in a separate radiograph, as in the bottom panel of **Figure 9**. However, these data by themselves are not

---

[47] The threshold value of Z>72 is used because elements between Z=57 (lanthanum) and Z=72 (hafnium), inclusive, are very rare in commerce, making 72 a reasonable boundary between high Z and lower Z elements for the purposes of CAARS.

[48] For a periodic table of the elements, see "WebElements" at http://www.webelements.com/.

sufficient to trigger an alarm. A typical cargo container has many overlapping objects, so another algorithm is needed to separate them from each other. Still another algorithm utilizes the foregoing data to calculate the size and Z of individual objects, and on that basis determines whether to trigger an alarm.[49]

High-energy x-rays could also exploit a characteristic of U-235 and Pu-239 (as well as of U-238 and thorium-232, non-threat materials that are relatively uncommon in commerce): they fission when struck by photons with an energy above approximately 5.6 MeV. As discussed in the **Appendix**, the resulting fission products decay over many seconds, producing prompt and delayed neutrons and gamma rays. High-energy x-rays may thus be used to detect high-Z material in general and SNM in particular. However, CAARS does not utilize this characteristic.

Technology description

This section focuses on the least complex and lowest-risk CAARS system, which is being developed by L-3 Communications Corporation. As **Figure 10** shows, it would use a concrete enclosure to minimize the radiation exclusion zone. The enclosure is 160 ft long in order to process two trucks at a time in 3 minutes so as to meet the CAARS throughput requirement of 40 containers per hour. The system would use one linear accelerator to generate 6-MeV electrons and (through the *bremsstrahlung* process) x-rays with energies from 0 to 6 MeV and, in the same manner, another accelerator to generate x-rays with energies from 0 to 9 MeV. A container would remain stationary as the beam is moved on a gantry over the truck. On the other side of the container are two detector arrays, one particularly sensitive to 6-MeV x-rays and the other to 9-MeV x-rays. Each array would record successive vertical slices of an opacity map of the container. The slices would be combined into a dual-energy radiograph as described above.

---

[49] For a technical discussion of dual-energy radiography, see S. Ogorodnikov and V. Petrunin, "Processing of Interlaced Images in 4B10 MeV Dual Energy Customs System for Material Recognition," *Physical Review Special Topics—Accelerators and Beams*, Volume 5, 104701 (2002).

Figure 10. L-3 CAARS Schematic Drawings

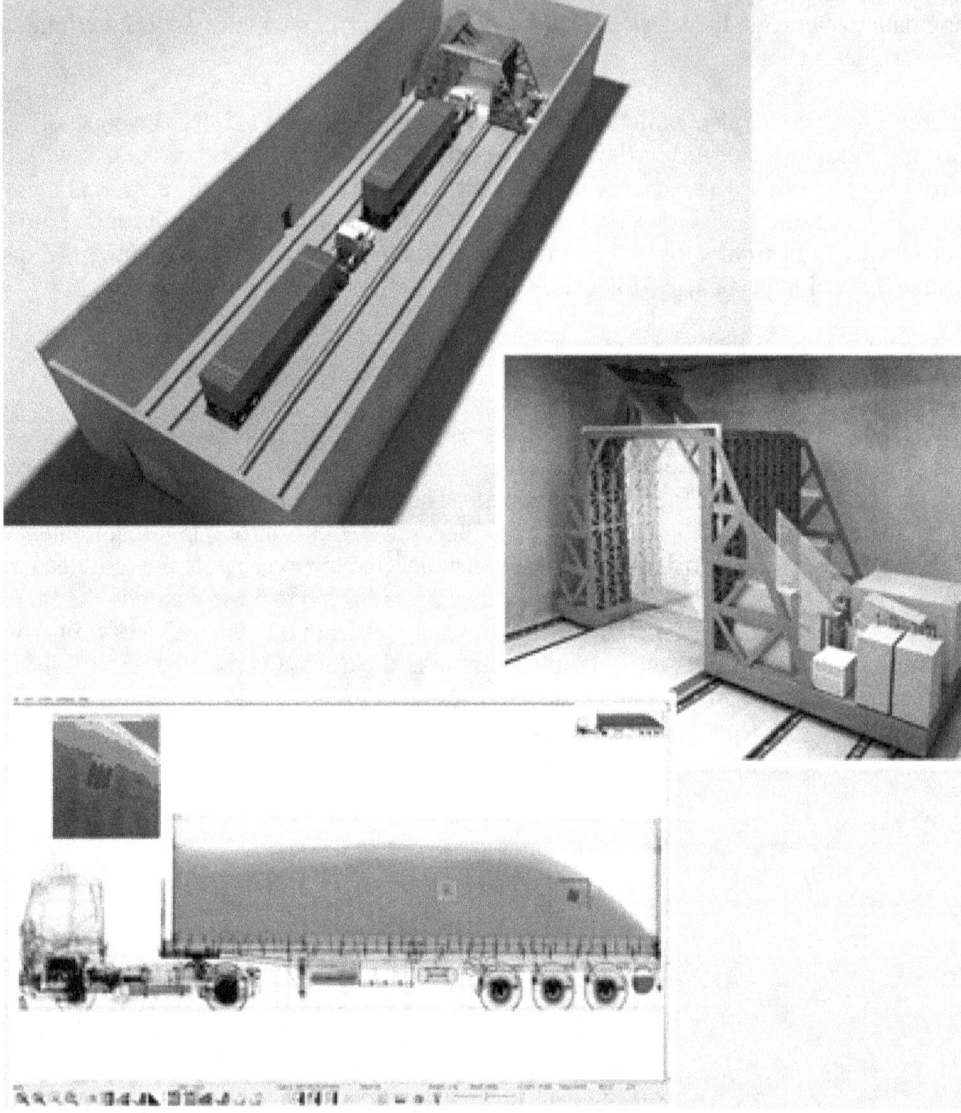

Source: Domestic Nuclear Detection Office.

Notes: The detection system is mounted on rails and scans by moving over the trucks. A concrete enclosure minimizes the radiation exclusion zone and thus the footprint. (middle) The unit has two accelerators (right) and two detector arrays (left). (bottom) This is a simulation of the image the system would return when it detects high-Z material in a cargo container.

Potential benefits

If CAARS works as DNDO anticipates, it could automatically detect high-Z materials while providing standard radiographs so that it would have potentially little or no impact on CBP operations, such as efforts to identify "traditional" contraband (drugs, guns, explosives). As such, CAARS could, if successful, integrate the SNM detection mission with CBP's historical mission of detecting other illegal materials. DNDO anticipates that the automated detection of high-Z materials would not slow down the pace of screening, which could continue to be determined by the rate at which operators can examine radiographic images for detection of normal contraband, though CBP expresses concerns on this point, as discussed below. CAARS technology is being designed to scan at least 40 40-foot containers an hour.[50] DNDO anticipates that it would have a 90% probability of detecting 100 cc of high-Z material (such as a cube 4.6 cm, or 1.8 inches, on a side), and a false alarm probability less than 3 percent, both with 95% confidence.[51] DNDO does not count detection of high-Z material as a false alarm because such material may be used to shield SNM. Another possible benefit of the CAARS program is that it is developing novel technologies and advancing the state of the art; even if no CAARS systems were to be deployed, scientific and technical advances made through this program could be of value to future detection systems. Comparing the L-3 and SAIC systems, the latter takes up less space, an important consideration for CBP at seaports where space is limited, but the latter has finer resolution, an important consideration for CBP in searching for traditional contraband.[52]

Status, schedule, and funding

In September 2008, Vayl Oxford, Director of DNDO, described developments with the CAARS program as follows:

> Consistent with any rigorous development and acquisition program, DNDO conducted system requirement reviews in November 2006 and preliminary design reviews in late May and June 2007 to assess the maturity of the CAARS technology. As a result, DNDO found that the technology was more difficult to implement than originally anticipated and determined that the technology should be demonstrated so that its full performance capability could be established prior to acquisition. It was also determined that the CAARS units, as currently designed, are too large and complex to be operationally effective. Finally, since 2006, there have been several technical advances in currently-deployed or soon-to-be-deployed NII systems that might provide some, but not all, of the desired capability. Accordingly, DNDO undertook a "course correction" in April 2008 and modified the three CAARS contracts to remove the "acquisition" component of the contracts, yet retain the demonstration and the test and evaluation (T&E) components of the contracts to allow collection of the required performance data.[53]

According to Joel Rynes, Program Manager, CAARS Program, DNDO, the L-3 CAARS project completed its Critical Design Review (CDR) milestone in July 2008.[54] The CDR marks the point at which the design has been completed and DNDO can give the contractor approval to begin fabrication of the prototype. As of February 2009, the prototype was being assembled in Las Vegas. It produced its first image in May 2009. The plan called for L-3 to collect data with the

---

[50] DNDO dropped its original goal of scanning 120 containers per hour. Testimony of Vayl Oxford, Director, Domestic Nuclear Detection Office, before the Senate Homeland Security and Governmental Affairs Committee, September 25, 2008.

[51] U.S. Department of Homeland Security. Domestic Nuclear Detection Office. "Cargo Advanced Automated Radiography (CAARS) Program Update Brief," presented to the Senate Committee on Homeland Security and Government [sic] Affairs, February 25, 2008, by William Hagan, Assistant Director, DNDO, Transformational and Applied Research, slide 4. This source is a briefing marked by the Domestic Nuclear Detection Office, the originating organization, "For Official Use Only." That office has approved CRS use of the information cited by this footnote.

[52] Personal communication, Joel Rynes, Program Manager, CAARS Program, DNDO, July 8, 2009.

[53] "Opening Statement of Mr. Vayl S. Oxford, Director, Domestic Nuclear Detection Office, Department of Homeland Security, before the Senate Homeland Security and Governmental Affairs Committee," September 25, 2008, p. 4.

[54] Information in this paragraph was provided by Joel Rynes, Program Manager, CAARS Program, DNDO, February 9, 2009.

---

system for four or five months to develop the algorithm for discriminating between high-Z and lower-Z material. DNDO conducted Technology Demonstration and Characterization (TD&C) in February and March 2010 to characterize the performance of the prototype.[55] The other CAARS programs are discussed in their respective sections below.

Part of the course correction was establishment of the Joint Integrated Non-Intrusive Inspection (JINII) Program—"joint" because the project is a collaboration between DNDO, CBP, and the DHS Directorate for Science and Technology, and "integrated" because it seeks to detect both traditional contraband and high-Z material. It is "non-intrusive inspection" (NII) in the sense CBP uses the term, namely, CBP personnel could clear containers, without physically having to open them, with high confidence that they do not contain contraband or SNM.

JINII has two main components. One is CAARS. The other is a test campaign by DNDO to examine the ability of existing, commercially available radiography systems to detect high-Z material by means of operators visually inspecting radiographs. This is distinct from CAARS, which is intended to highlight suspicious areas automatically. In addition, systems have been developed outside of the CAARS program that have a limited capability—which DNDO anticipates would be less than that of CAARS—to detect high-Z objects in cargo automatically. One of these systems, the Rapiscan Eagle Portal, completed TD&C in September 2009.[56] DNDO states that if tests demonstrate this limited capability with commercial off-the-shelf systems, deploying such systems could place capability in the field sooner than would be the case by using only CAARS systems.

CBP already deploys the Rapiscan Eagle, which uses an accelerator (6 MeV in some versions, 4 MeV in others) to generate a radiographic image of a cargo container. The version that was tested through JINII has an added algorithm, "Auto-Z," that is designed to detect high-Z material in cargo and indicate the location of such material on a radiograph automatically. In contrast to either CAARS candidate, it would cost less, would use the current supporting infrastructure, and would require little added operator training. According to Rynes, the algorithm should be able to detect a 400-cc cube of high-Z material (7.37 cm, or 2.9 inches, on a side) nearly as well as CAARS candidates, but could not detect a 100-cc cube of high-Z material (4.65 cm, or 1.8 inches, on a side) as well as they could.[57]

In March 2010, DNDO completed its Technology Demonstration and Characterization (TD&C) testing for the L-3, Rapiscan, and SAIC systems and put the CAARS follow-on program on hold. DNDO expects the TD&C to provide data to quantify the performance and cost of these three systems so that it can perform a cost-benefit analysis that would consider both high-Z and traditional contraband detection, helping it decide how to proceed. DNDO could recommend future development, operational testing, acquisition, or some combination, of the various systems. If DNDO, CBP, and Congress judge the L-3 or SAIC systems to be more cost-effective than existing systems, one or more systems with CAARS-level capability could be further developed

---

[55] "Developmental Test and Evaluation" (DT&E) and "Technology Demonstration and Characterization" (TD&C) are both tests of systems conducted by the government (as opposed to contractors). DT&E is a term used in the acquisition process; DT&E tests systems before beginning procurement. As noted earlier, DNDO undertook a "course correction" in April 2008, moving CAARS from an acquisition program to an R&D program. Some felt it was confusing to use the term DT&E for a non-acquisition program, so DNDO decided to call the government tests of CAARS candidates TD&C.

[56] Information provided by Joel Rynes, April 16, 2010.

[57] Information provided by Joel Rynes, July 8, 2009.

or acquired through a competitive process yet to be determined. With TD&C completed, the status of the test equipment as of May 2010 was as follows, according to Rynes,:

> Significant portions of the AS&E CAARS material have been distributed to other DNDO projects. The SAIC CAARS system is being dismantled and significant portions of the material are being distributed to other DNDO projects. The L-3 CAARS system is being dismantled and stored for potential installation at a later date. This would be [done] outside of the CAARS program. The Rapiscan Eagle remains installed at Moffett Field [CA]. It has funding to stay there through FY11.[58]

The budget for the entire CAARS program is: FY2006, $16.3 million; FY2007, $26.4 million; FY2008, $31.8 million; and FY2009, $26.1 million. The CAARS program per se ended in FY2009, to be replaced by a followon DNDO-CBP program to advance CAARS technology so that it can be deployed in the field. FY2010 funding for this program is $15.2 million.

Risks and concerns

Scientific risks and concerns

The L-3 system is intended to be the least complex and lowest-risk CAARS system. The scientific risk to its hardware is low because it uses commercially available accelerators and detectors. The software risk is potentially higher because algorithms to sort high-Z and low-Z materials on a radiograph automatically have not been fully developed and have only been modeled in a simulated environment, not tested in an actual operational one. The other two CAARS technologies are more complicated, so their scientific risk may be greater.

Engineering risks and concerns

The concept has been demonstrated in the laboratory, but it remains to be shown that it can be scaled up to the size CBP needs to scan containers. For example, can the algorithm to sort pixels into high-Z and lower-Z bins handle large enough quantities of data in a timely manner? At high enough energy levels, an accelerator can produce neutrons that require a large amount of shielding. The x-rays generated by the accelerator can also scatter from interactions with the cargo, also requiring shielding. A large footprint for shielding could preclude deployment of detection systems at some ports. At issue: Can radiation be kept low enough that the footprint does not become excessively large?

Cost and schedule risks and concerns

Slipping the CAARS schedule through the "course correction" to permit more time for R&D could make it easier to meet the new schedule, reducing schedule risk. More R&D could also reduce cost risk. On the other hand, scientific risk could result in cost and schedule risk.

Another cost and schedule risk is that work on three projects would reduce effort devoted to any one system, delaying each system (as compared to the schedule possible if the full funding had been applied to only one system) and increasing the cost of the total program. For example, it

---

[58] Personal communication, e-mail, May 12, 2010.

could be argued that a more efficient use of funds would be to focus only on the L-3 system because its development is furthest along. On the other hand, a multi-pronged approach may offer countervailing advantages. The simpler technology could (presumably) be deployed quickly, adding capability quickly and providing a hedge against failure of more advanced systems. More-capable systems could be deployed later, reducing the time pressure to develop them. Conversely, pursuing several approaches hedges against the prospect that the simpler system might encounter unanticipated problems that delay it to the point where it and a more advanced system could be deployed at about the same time. Whether these advantages justify the higher cost is always a matter of debate.

CBP notes that other companies are developing systems to detect high-Z material outside of the CAARS program, and argues that, because of the operational concerns discussed next, the money spent on CAARS could be better spent on such systems.[59] DNDO points out that the JINII program is developing and evaluating some of these systems.[60]

Operational risks and concerns

(1) A 3% false alarm rate, if achieved, would place a large burden on CBP CONOPS, as it would require responding to many false alarms. Alternatively, as discussed in the section on computer modeling, a high false alarm rate could lead operators to ignore alarms or to raise the threshold for alarms, increasing the likelihood of missing an actual threat. (2) Joel Rynes, Program Manager, CAARS Program, DNDO, states that an operator could clear some false alarms without intrusive inspection, such as by inspection of radiography images, reducing the rate of alarms requiring intrusive inspections to well below 3%. At issue is whether there is enough confidence in operator judgment with these methods to avoid intrusive inspections. (3) The system does not differentiate between SNM and other high-Z material; adding non-SNM high-Z alarms to false alarms boosts the non-SNM alarm rate and, presumably, the rate of alarms requiring intrusive inspections. Without a way to differentiate between SNM and other high-Z material, can CAARS meet its goal of having "little or no impact on CBP operations"? On the other hand, is any high-Z material suspicious and worthy of inspection, so that non-SNM high-Z alarms should not be counted as false alarms? (4) A CAARS program goal is automated detection of small amounts (100 cc) of high-Z material through 10 inches of steel, with a follow-on goal of penetrating 16 inches of steel.[61] Is that a reasonable goal? If not, what is the maximum thickness of steel commonly found in cargo containers and what energy level of photons would be required to penetrate it? How would the added shielding required by equipment that generates photons greater than 9 MeV affect deployment at ports, where space is at a premium? Or would no practical energy level suffice to penetrate that much steel? Alternatively, would a simple "Bucky collimator," as described under "Background" in this section, make radiography much more effective? (5) The radiation produced by CAARS, or other high energy imaging systems, even at low levels, may cause interference with radiation portal monitors that have already been installed.

---

[59] Information provided by Ira Reese, Executive Director, Laboratories and Scientific Services, and Patrick Simmons, Director, Non-Intrusive Interrogation, both of Customs and Border Protection, personal communication, October 23, 2008.

[60] Information provided by Joel Rynes, personal communication, October 28, 2008.

[61] Domestic Nuclear Detection Office. "Cargo Advanced Automated Radiography (CAARS) Program Update Brief," slides 3, 18. This source is a briefing marked by the Domestic Nuclear Detection Office, the originating organization, "For Official Use Only." That office has approved CRS use of the information cited by this footnote.

It is important to minimize this interference, though it can readily be eliminated by turning off the accelerator when portal monitors are in use.

CBP raises other operational concerns. The first concern applies to the L-3 and AS&E CAARS systems; the others apply to all three. (1) Because of an apparent miscommunication between DNDO and CBP, DNDO thought that the dimensions of a one-of-a-kind CBP radiography system, 60 x 160 ft, were acceptable to CBP for large-scale acquisition, some 300 to 500 units nationwide. The L-3 and AS&E CAARS systems would use a concrete enclosure of that size (see **Figure 10**) for radiation containment. However, CBP states that no unit that size would fit in U.S. seaports, and that only four or five ports of entry on the U.S.-Mexican border could accommodate the units. (2) CBP expresses concerns that the radiation emitted by radiography equipment such as CAARS could require a large exclusion zone to protect workers. In contrast, radiation portal monitors emit no radiation, but passively sense radiation emitted by radioactive material. (3) CBP is concerned that total scan time would create an immense delay for container traffic entering the United States from seaports or land border crossings. While DNDO's goal is to have CAARS scan a container in 15 seconds, CBP notes that the time when the scanning unit is on is only a small part of the total time a scan requires. It points to the following sequence: a driver would pull a truck with a container into the scanning enclosure (depicted in **Figure 10**); the driver would leave the truck and go to a radiation-protected facility; the scanning equipment would move over the stationary container at a precisely controlled speed; the truck and container would be scanned; and the driver would reenter the truck and drive it out of the enclosure. This sequence, CBP estimates, could take 5 minutes.[62]

Thomas Winkowski, Assistant Commissioner, Office of Field Operations, CBP, provided was emphatic on the need to minimize delays. Not referring specifically to CAARS, he said that rebooting and reinstalling software on some systems "can take seven or eight minutes. That's the kiss of death in my business from the standpoint of delays." Regarding CAARS, he said that CBP was "at the table" with DNDO in discussing requirements.

> But what our concern was was that the footprint was too big ... and the throughput, you know, for all your trucks to go through a car wash type system, as I call it, and the driver comes out, and you do your scan—that realistically presents a tremendous amount of problems from cycle time. So our position was that we really needed a different technology that was more flexible and that didn't have such a big footprint and require so much handling.[63]

Rynes stated that concerns such as the foregoing raised by CBP were among the reasons for the CAARS course correction and the establishment of the JINII program.[64]

Potential gains by increased funding

Rynes states that added funds could improve CAARS in several ways: (1) Further development could reduce the size of CAARS systems so they would be more readily deployable at ports. (2)

---

[62] Ira Reese, Executive Director, Laboratories and Scientific Services, and Patrick Simmons, Director, Non-Intrusive Interrogation, both of Customs and Border Protection, provided information in this paragraph, personal communication, October 23, 2008.

[63] Testimony of Thomas Winkowski, Assistant Commissioner, Office of Field Operations, U.S. Customs and Border Protection, before the Senate Homeland Security and Governmental Affairs Committee, September 25, 2008.

[64] Information provided by Joel Rynes, personal communication, October 28, 2008.

Money spent (by CAARS or another project) to improve detector efficiency could permit a less-powerful electron source to suffice, further reducing the requirements for shielding. (3) Added funds could improve algorithms to automate the processing of radiographs. (4) JINII is just beginning to develop algorithms to detect contraband automatically to increase the rate at which containers are inspected. Added funds would accelerate this development. (5) Added funds could expedite better integration and data fusion of CAARS with already-deployed radiation portal monitors.[65]

Potential synergisms and related applications

The preceding paragraph discussed synergisms by which technology developments could improve CAARS capabilities. Another synergism concerns the use of neutrons or high-energy photons to induce fission to discriminate SNM from other high-Z material. DNDO issued a broad agency announcement in March 2008 to develop this capability.[66] [67] The CAARS candidates do not include technology for this purpose. For example, they do not count neutrons or gamma rays that would be generated by photofission even though 9-MeV photons can cause that effect. However, CAARS might possibly draw on such work in the future.

SAIC CAARS: A Higher-Risk, Higher-Benefit Dual-Energy Radiography System[68]

The "problem" and "background" information for the L-3 system, as described above, are the same as for the SAIC system.

Technology description

Science Applications International Corporation (SAIC) developed a system (unnamed) as part of the CAARS program. That program completed Technology Demonstration and Characterization of the competing systems in March 2010, and as of April 2010 the future of CAARS was unclear. However, SAIC is competing to apply the technology it developed under CAARS to another DHS program as described below.

SAIC's basic CAARS system uses an electron accelerator developed by Accuray Corporation.[69] The accelerator is "interleaved"—a single unit produces pulses of electron beams alternating between 6 and 9 MeV. Each pulse lasts 3 microseconds, with 2.5 milliseconds between pulses.

---

[65] Personal communication, July 11, 2008.

[66] U.S. Department of Homeland Security. "Advanced Technology Demonstration for Shielded Nuclear Alarm Resolution." Broad Agency Announcement 08-102 for the Domestic Nuclear Detection Office, Transformational and Applied Research Directorate, March 2008, p. 22, available at https://www.fbo.gov/files/6f6/6f6f82cc03fe6fcbf298a7e3903a15b7.doc?i=640d9818cdf76fa884b8c064b8f19376.

[67] As of April 2010, all Shielded Nuclear Alarm Resolution contractors had completed their preliminary design reviews by September 2009 and are working toward their critical design reviews, in which DNDO would approve the design. The first unit is scheduled to begin testing early in CY2011. Information provided by Joel Rynes, April 16, 2010.

[68] Rex Richardson, Vice President and Principal Scientist, Science Applications International Corporation, provided detailed information for this section, personal communications, June-August 2008. Others have commented as well to provide alternative perspectives.

[69] See http://www.accuray.com/.

These beams strike a copper target and generate photons with a spectrum of energies up to the highest energy of the electron beam, though lower-energy photons are filtered out.[70] During a scan, the photons pass through a cargo container in a vertical fan-like beam. On the other side of the container is a detector array composed of multiple detector elements arranged in thin vertical bands. Each element records an opacity image of a narrow horizontal "stripe" of the container. An algorithm combines the stripes into a dual-energy radiograph, as described in the L-3 section. The system automatically flags for further inspection areas of high-Z material and areas with too much dense material for the photons to penetrate. **Figure 11** illustrates the configuration of a unit.

Figure 11. SAIC CAARS Prototype

Source: Photograph provided by SAIC, May 2010

Notes: A bridge-like structure is mounted on rails some 20 ft apart. The unit on the far left houses the dual-energy accelerator and electronics. The vertical housing to the right of the truck is in line with x-rays generated by the accelerator. It contains detectors that record the radiograph, pixel by pixel. To the right of the housing is a steel plate 12 ft high by 20 ft wide by 1 to 7 in thick that absorbs x-rays that scatter off the cargo in the direction of the beam. It weighs some 70,000 lb but is part of the unit, so the unit is self-contained. A thinner plate with the same length and width to the left of the truck stops lower-energy photons that scatter backward

---

[70] Electron beams of the levels discussed here generate *bremsstrahlung* photons of a wide spectrum of energies. Most of the photons are of relatively low energy, well below 1 MeV. They contribute almost nothing to the radiographic image but can represent a major portion of the radiation exposure to personnel. To remove or "filter" them, a piece of copper is placed in front of the photon beam. The material is thick enough to stop lower-energy photons but not higher-energy ones.

from the cargo. In operation, the driver exits the truck, and the structure—including accelerator, detector, and steel plates—moves over the truck at 3 ft/sec. The control booth on the right is stationary; the tower on the far right is not part of the system. This photograph was taken at SAIC's San Diego factory in September 2009 during readiness testing in preparation for DNDO's Technology Demonstration and Characterization event.

One key to this system is that the proprietary detector operates at a photon flux (number of photons per unit time) one-hundredth that of conventional cargo imaging systems. A lower flux enables the accelerator to be much more compact. Accelerators with high photon flux require a high-Z material, typically tungsten, for the *bremsstrahlung* target because it can withstand the high heat from the electron beam. Such accelerators also use high-Z material to shield the photon beam. However, photons with energies greater than approximately 6 MeV produce photoneutrons (neutrons knocked off atoms by high-energy photons) when they strike high-Z materials, with the number of such neutrons increasing as energy increases. The heat generated by the beam with lower photon flux is low enough that copper can be used instead of tungsten. Copper is one of a few metals with a threshold greater than 9 MeV for producing photoneutrons. As a result, SAIC states, its system produces virtually no photoneutrons, eliminating the need for large concrete structures for neutron shielding. Further, since the system has a low beam flux, SAIC states that the need for shielding from x-rays that scatter in the cargo is greatly reduced.[71]

Another key to the system is the dual-energy interleaved accelerator. Three aspects are especially consequential. First is the ability to construct this type of accelerator; previous efforts by Varian Medical Systems, funded by SAIC in 2004, had poor X-ray beam stability.[72] Second, this accelerator uses a lower beam flux, with the advantages just discussed. Third, the number of photons per pulse ("dose repeatability") varies by a very small amount, less than 0.4%, far below the required variance of less than 5 %.

While the goal of CAARS is to find high-Z material while not interfering with efforts by CBP inspectors to find traditional contraband (drugs, guns, money, etc.), the better-than-expected dose repeatability enabled the system to differentiate approximately 15 bands of Z from carbon (Z=6) to uranium (Z=92), an atomic number resolution of about six (e.g., the difference between oxygen, Z=8, and silicon, Z=14). This contrasts with the ability to find only materials with Z greater than 72, as in **Figure 9**. On a radiograph in various shades of gray, the difference may not be apparent at all, but if each Z band is represented by a different color, different materials display much more clearly, as **Figure 12** shows. Colors are assigned to Z bands arbitrarily, producing a "false color" or "pseudocolor" radiograph, so called because the colors bear no relationship to the color of materials being radiographed.

How does reducing the variance increase the number of Z-bands? Dual-energy radiography finds the thickness, as measured by each of the two beams, of the material recorded for each pixel. For example, it finds the number of photons (in this case, x-rays) transmitted through a cargo container, and then received at the detector that measures the number of photons at each pixel. An algorithm calculates the ratio of the two thicknesses and displays this ratio as an image in colors or grayscale, pixel by pixel. Thickness ratios are calibrated before each scan. Because the number

---

[71] The beam flux is further reduced by filtering. Electron beams of the levels discussed here generate *bremsstrahlung* photons of a wide spectrum of energies. Most of the photons are of relatively low energy, well below 1 MeV. They contribute almost nothing to the radiographic image but can represent a major portion of the radiation exposure to personnel. To remove or "filter" them, a piece of copper is placed in front of the photon beam. The material is thick enough to stop lower-energy photons but not higher-energy ones. After filtering, 6- and 9-MeV beams generate the greatest number of photons at about 2 and 3 MeV, respectively.

[72] Beam stability refers to uniformity in both the duration of an x-ray pulse and the number of x-rays in a pulse.

of x-rays varies from pulse to pulse, the output end of the accelerator has an x-ray dose meter to correct for this variance. However, this correction can introduce large errors in the ratios measured at some pixels because the correction applies only to the beam output and not to the beam as received at different points. If there were no variance at all in the output (number of x-rays per 6-MeV beam and per 9-MeV beam), this correction would not be needed, eliminating this error source and increasing the number of Z bands that can be displayed. At the same time, even with zero variance, the number of Z-bands that can be displayed is limited, and can be reduced, by such factors as the size and thickness of objects being radiographed, and by the mix of objects of different Z numbers that a beam might pass through for a given pixel.

Figure 12. Dual-Energy Radiography with False Colors Assigned

Bottom Image Can Differentiate Between Approximately15 Bands of Atomic Numbers (Z)

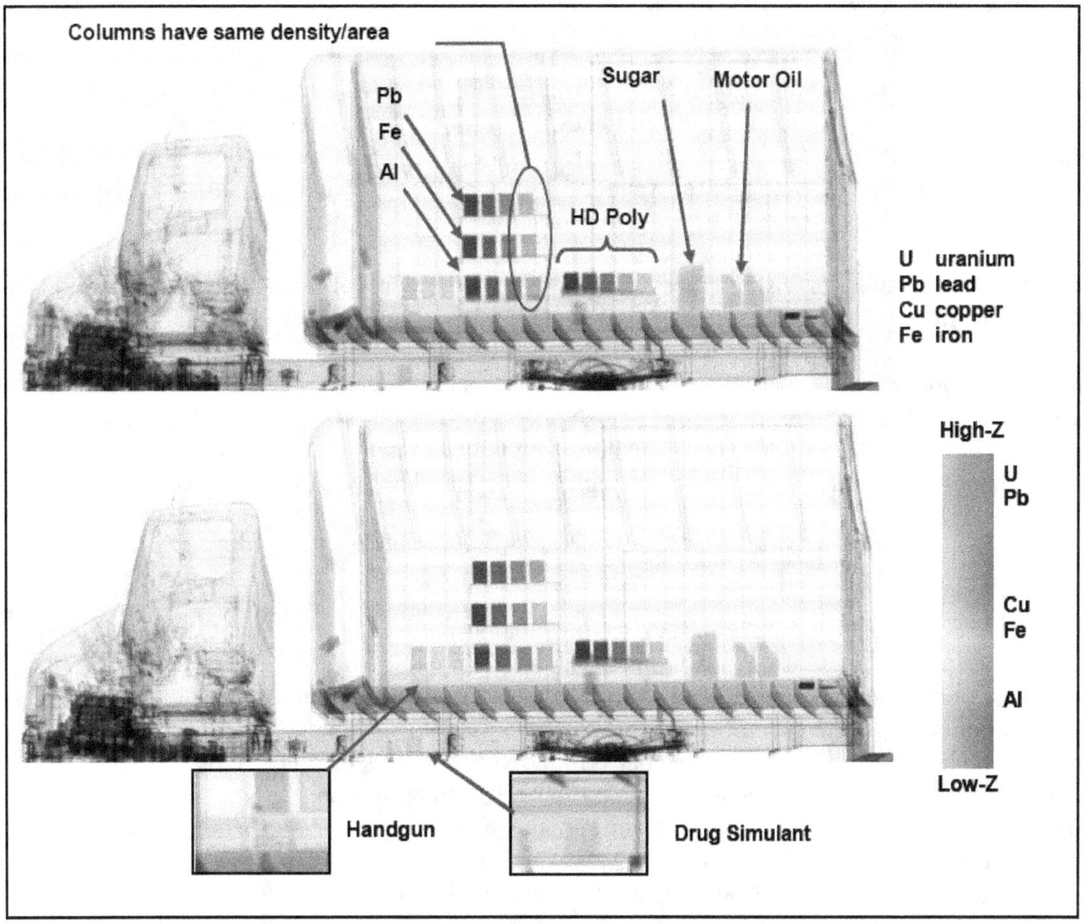

SAIC CAARS* 6 MeV / 9 MeV dual energy separation of materials by atomic number
(*work funded by DHS DNDO)

Source: Image provided by SAIC, April 2010.

Notes: This figure shows two radiographs of the same objects in a HINO delivery truck. The top image is a conventional radiograph made using the 9 MeV beam. The bottom image assigns a color arbitrarily to each pixel according to the ratio of (a) the thickness as measured by the 9-MeV beam to (b) the thickness as measured by the 6-MeV beam. Higher ratios (higher Z) are displayed in this image as being toward the red end of the spectrum while lower ratios (lower Z) are displayed more toward the blue end, though the algorithm can display selected Z bands in whatever color the operator chooses so as to facilitate searching for items of concern for a particular cargo container.. As a result, the colors bear no physical relationship to the material being radiographed. Because the 6- and 9-MeV beams have little variance in the number of photons per pulse, it is possible to maintain a precise calibration of the relative intensities, leading to the 15-band atomic number resolution. An operator could use the colors not only to find high-Z material, but also to find contraband hidden in a material of slightly different Z. The algorithm can highlight anomalous areas, in this case a handgun and a drug stimulant.

Text reading "columns have the same density/area" refers to the four stacks, or columns, in the middle of each image. For each stack, the three blocks were selected to have the same radiographic density. That is, in each stack, the thickness of each block was varied so as to have the same gray-scale intensity (i.e., to allow the same number of photons to pass through per unit time) as the other blocks in that stack. The goal was to show that two (or more) materials could look exactly the same in a simple gray-scale radiograph (top image) but could be distinguished as different materials by false-color imaging constructed using data from the dual-energy scan. Conversely, the five blocks labeled "HD poly" were selected to show that the same material can have different gray-scale intensities depending on thickness, but the false-color imaging reveals them to have the same color even though the intensity of the color varies with thickness.

Abbreviations: Z, atomic number; HD poly, high-density polyethylene; U, uranium; Pb, lead; Cu, copper; Fe, iron; Al, aluminum; DHS, Department of Homeland Security; DNDO, Domestic Nuclear Detection Office

Two of CBP's concerns with equipment to detect possible terrorist nuclear weapons or fissile material are that that mission is added on top of the mission of detecting traditional contraband, adding to the workload of CBP front-line operators, and that the equipment does not help detect contraband. Yet in contrast to nuclear weapons or material, contraband is a constant threat, with criminals attempting to smuggle in many tons of it every day, and often succeeding.

The ability to separate pixels into many bands by Z would address these concerns, as it would help an operator find typical contraband hidden in a cargo container as well as high-Z material. Such contraband has a Z of less than 72, so the ability to find high-Z material does not contribute to finding contraband. In contrast, dividing pixels into many Z-bands facilitates the detection of anomalies, such as guns (Z~26) hidden in a shipment of water (Z~10). Further, the current software enables the operator to choose colors with which to tag different Z-bands in order to make shapes stand out. Algorithms to highlight suspicious shapes could also be developed.

Potential benefits

Greatly reducing neutron flux offers several potential advantages. (1) Worker exposure to neutrons is reduced. (2) Lower flux requires less shielding, thereby reducing cost and footprint. (3) If the basic system is augmented by the capability to detect neutrons and gamma rays resulting from photofission, neutron detectors in this system would be able to detect neutrons when the beam is on because of the low neutron background. SAIC claims that its CAARS system is the only one with this capability. Since most neutrons released by fission of SNM are "prompt" (released immediately), the prompt neutron signal is much larger, and thus easier to detect, than that of delayed neutrons. Fission of SNM generates high-energy neutrons; by adding detectors that can discriminate between neutrons on the basis of their energies, SAIC expects that its system will be able to determine if a high-Z object is SNM. (4) The accelerator is very compact, 80 cm in length, also reducing footprint. (5) The system is designed to flag high-Z material automatically, reducing the burden on operators and the risk of operator error. Another benefit of the system is that it occupies considerably less space than do the L-3 and AS&E systems, an important factor for locations, such as seaports, where space is at a premium.

Status, schedule, and funding

SAIC is developing its basic system under contract to DNDO. According to Rynes, the SAIC system has been making substantial progress since its Critical Design Review in April 2008, as follows.[73] The system has been built and has been collecting images. (Images are what an operator sees.) As of February 2009, it was collecting data for developing the data-processing algorithm. As of July 2009, a test unit had been built and SAIC was refining its algorithms. Technology Demonstration and Characterization was completed in December 2009. The Accuray interleaved x-ray source is the key to making the overall system smaller, a major criterion for CBP. Rex Richardson, Vice President and Principal Scientist, SAIC, said in May 2009, "We have resolved all of the initial start-up issues related to the Accuray compact X-ray source and the source is now performing beyond expectation. Hence I think I am confident in saying that the 'higher risk' aspect of the SAIC program has now been resolved and we are collecting the 'higher benefit.'"[74] He stated in February 2009 that SAIC is "at the near-production prototype stage and can produce pilot test units for deployment at ports and border crossings in a few months" and that SAIC is "now imaging full cargo loads at our design speed of 33 inches per second."[75]

In February 2009, SAIC estimated unit price of its system at $4 million to $7 million, depending on terms of procurement such as number of units ordered, delivery schedule, and warranty agreements.[76] In September 2008, DNDO did not elect to fund SAIC's proposal for an advanced technology demonstration of the add-on capability to its CAARS system discussed earlier, to detect shielded SNM by detecting radiation released by fission of SNM induced by high-energy x-rays.

As of April 2010, the status of the SAIC program was as follows. Without funds to continue its program, SAIC disassembled its CAARS test unit and was disposing of the government-owned material under DNDO supervision. Meanwhile, SAIC was adapting its technology to a truck-mounted design in response to Broad Agency Announcement 10, Non-Intrusive Inspection and Automated Target Recognition Technologies, or "CanScan," issued by the DHS Directorate for Science and Technology.[77] One element of CanScan supports development of a next-generation mobile cargo imaging system for CBP operators to use at seaports and border points of entry. If it is awarded a contract under CanScan, SAIC anticipates that it would develop a prototype that would integrate its CAARS dual-energy technology with such techniques as neutron active interrogation for materials identification, and that it would deliver a production-ready prototype truck platform to CBP within three years of contract award.

---

[73] Personal communication, February 9, 2009.

[74] E-mail from Rex Richardson, May 11, 2009.

[75] E-mail from Rex Richardson, February 9, 2009.

[76] E-mail from Rex Richardson, SAIC, February 17, 2009.

[77] U.S. Department of Homeland Security. Science & Technology Directorate. "Non-Intrusive Inspection and Automated Target Recognition Technologies," Broad Agency Announcement BAA-10, CanScan, December 16, 2009, http://www.aqd nbc.gov/Business/uploads/BAA10-CanScan.pdf.

Risks and concerns

Scientific risks and concerns

This system required completing the interleaved accelerator, yet its development was challenging. As a result, it involved more scientific risk than the L-3 system, which uses two separate accelerators, one for each energy level. An earlier attempt to develop an interleaved accelerator encountered a problem with the stability of the electron beam, with energies varying by a factor of two. The SAIC CAARS system requires that the Accuray accelerator demonstrate that it can repeatedly generate electrons of two energy levels, each within a narrow energy band. As of May 2009, the accelerator was operating beyond expectations, greatly reducing if not eliminating its development as a source of technical risk. As noted, accelerator continued to operate beyond expectations, which among other things enabled the system to complete Technology Demonstration and Characterization in December 2009.

Engineering risks and concerns

Typical cargo imaging systems have a resolution of 3 mm to 5 mm (i.e., they can display details that are 0.12 to 0.20 inch in any dimension). The design of the SAIC system results in resolution of 7 mm to 9 mm. DNDO has set a standard of detecting 100 cc of SNM (e.g., a cube 4.6 cm on a side), so the significance of this loss of resolution is not clear for SNM detection, though it would probably reduce the system's ability to detect other contraband. CBP prefers finer resolution to help spot contraband. SAIC responds that it can achieve 5-mm resolution by using more detectors, albeit at higher cost; that such fine resolution is not absolutely required for scanning cargo containers; and that false-color imaging would have great value for discerning contraband. Another concern is the extent to which different thicknesses of materials or different materials together in a container would reduce the number of Z-bands that the system can display.

Cost and schedule risks and concerns

(1) CBP insists that scanning should interfere with the flow of commerce as little as possible. In response, DNDO requires the system to scan containers at a rate of 2.7 ft per second, the speed needed to scan a 40.5-ft container in 15 seconds. (This time refers to the actual time when a container is being scanned, as distinct from the requirement to process 40 containers per hour, which includes time for a container to move to and from the scanning equipment.) The 15-second requirement imposes a burden on the system. Improving the performance of the system at a given scan speed and photon output would require a significant redesign and additional cost. On the other hand, some ask, since radiographs of containers must still be scanned visually by operators to search for contraband, which may take more than 15 seconds, is a 15-second scan time needed, or could that requirement be relaxed? (2) The system uses an interleaved accelerator that has proven to be technically feasible, but there is as yet no assurance that it will be available at the quantity, schedule, and cost needed to make the system competitive. (3) At a unit price of perhaps $5 million or more, some ask, is the system too costly to deploy in large numbers?

Operational risks and concerns

Uncertainties about the final configuration of the deployed system could affect operations. Space is at a premium in many U.S. and foreign ports. If a more powerful accelerator is needed in order to obtain finer resolution, it could require more space, more shielding, and more standoff distance, and might increase concern among port workers about radiation exposure. An alternate means of obtaining finer resolution would be to add more detectors, which would increase cost but not radiation.

Potential gains by increased funding

SAIC states that with added funds it could (1) integrate detection of photofission with CAARS radiography, (2) hire more software engineers to speed development of algorithms to improve the system's ability to differentiate between materials; and (3) pursue application of the system to detect contraband. Added funding would also allow SAIC to apply its CAARS technology to the CanScan program.

Potential synergisms and related applications

SAIC believes that there is significant potential for its dual-energy radiography technology to help detect contraband and explosives because it can differentiate between organic materials, which have a low Z, and most metals. Improved scintillator material being developed by several teams of scientists might be of use to the SAIC system. Similarly, algorithms being developed for other gamma-ray detectors or radiography units might have elements similar to those being developed by SAIC.

AS&E CAARS: Using Backscattered X-Rays to Detect Dense Material[78]

This system addresses the same problem as the L-3 and SAIC CAARS.

Note: On March 10, 2009, DNDO terminated the contract with American Science and Engineering, Inc. (AS&E) to continue developing this system.[79] DNDO views the technology incorporated in this system as holding some promise, but states that development of this technology requires additional basic research.[80] Accordingly, this section will not be updated further.

Background

AS&E is developing a CAARS system that would utilize a different physical principle than the L-3 and SAIC systems. Its core technology is "EZ-3D^{TM}," developed by Passport Systems, Inc. The term is an abbreviation for "effective" (i.e., average or approximate) Z (atomic number) of the material being detected, located in three dimensions. It is intended to exploit the principle that

---

[78] Jeffrey Illig, Program Manager, AS&E CAARS Program, and Stephen Korbly, Director of Science, Passport Systems, provided detailed information for this section, October 2008. Others have commented as well to provide alternative perspectives.

[79] U.S. Securities and Exchange Commission, *Form 8-K, American Science and Engineering, Inc.*, Washington, DC, March 10, 2009, p. 2, http://www.sec.gov/Archives/edgar/data/5768/000110465909017589/a09-7688_18k.htm.

[80] Information provided by Joel Rynes, personal communication, July 8, 2009.

when x-rays are beamed at an object, the number of x-ray photons that scatter backwards (in the opposite direction from the beam) is strongly proportional to the object's Z.

When x-rays strike high-Z material, they knock pairs of electrons and positrons off atoms. (Positrons are electrons with a positive charge.) These electrons and positrons travel in all directions. When they strike other atoms, they create *bremsstrahlung* photons (x-rays) that also travel in all directions. By arranging photon detectors so that they detect x-rays scattered at a backwards angle (>90 degrees) from the beam, they could detect these x-rays and not x-rays from the beam. The number of these backscattered x-rays and their energy distribution (as shown in **Figure 13**) is an indicator of Z. In an ideal situation—for pure chemical elements, and with no intervening material—the number of x-rays is approximately proportional to Z to the fourth power, so that number is enormously higher for high-Z material than for medium-Z material. In the real world, most goods shipped (e.g., wood, steel, plastic, electronics) are not pure elements and differ in size and shape, and a container may hold mixed types of cargo. As a result, experiments have shown, the effect is somewhat less. The effect of Z on number and energy of backscattered x-rays is the key to EZ-3D; **Figure 13** shows, for five elements, the difference in backscattered spectra as Z increases.

Figure 13. EZ-3D™ Differentiates Between Elements by Atomic Number (Z)

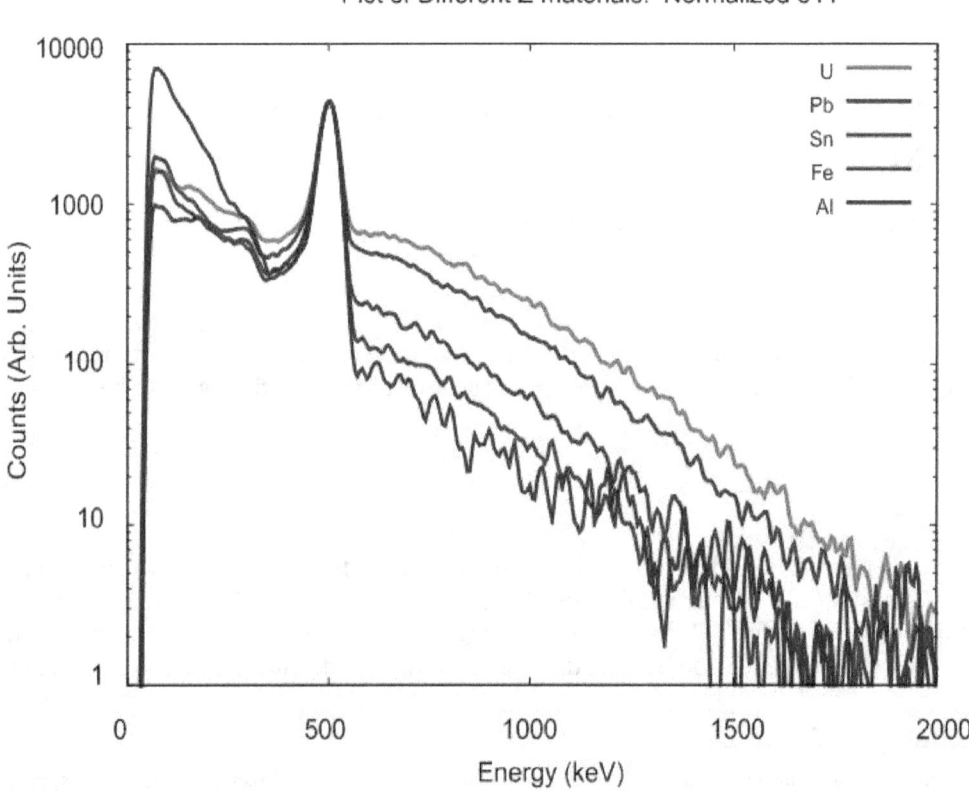

Source: Passport Systems, Inc.

This graph shows the number of backscattered photons (vertical scale) at each energy level (horizontal scale) recorded by a detector. The vertical scale is logarithmic, so vertical increments are larger than they appear. The plots are for five elements, from top to bottom, with Z: U, uranium, 92; Pb, lead, 82; Sn, tin, 50; Fe, iron, 26; Al, aluminum, 13. "Normalized 511" means that the graphs are "normalized" at 511 keV. That is, for each element, the number of counts at 511 keV is multiplied by the factor needed to set the count to a specified number (the same number is used for all elements in a scan) and counts at each other energy level for that element are multiplied by the same factor. This facilitates comparison across elements. For the region between roughly 600 keV and 1300 keV, the graph shows a difference between uranium and lead, and a larger difference between lead and tin. The distribution of points in the graph depends on the energy of the electron beam used to generate the x-rays via the *bremsstrahlung* process. For this graph, the beam has an energy of 2.8 MeV. Since almost all the x-rays have energies far below that level, with the peak number of x-ray photons at 300 keV, the number of counts diminishes at higher energy levels. For this graph, beyond about 1300-1800 keV, depending on the element, the number of background counts exceeds the number of backscattered counts, so that counts at and beyond those levels provide no useful data.

Technology description

Stephen Korbly, Director of Science at Passport Systems, Inc., describes the proposed design and operation as follows. A truck would pull a cargo container through the inspection unit at slow speed, 2.5 feet per second (1.7 mph). The container would first pass through a radiography unit to identify areas of dense cargo for the EZ-3D beam to examine in more detail.[81] The container next would pass through the EZ-3D unit. There, a Rhodotron[82] would generate a 9-MeV electron beam. This beam would travel through a series of electromagnets so as to steer the beam downward toward the container. The beam is designed to move back and forth transversely across the top of the cargo container to interrogate a "slice" of the container, as shown in **Figure 14**. If the system were to detect a volume of dense cargo, the beam could dwell on that volume longer to gather more data.

The electron beam would strike a water-cooled metal target, producing a spray of x-rays through the *bremsstrahlung* process. The x-rays would pass through a metal sheet that filters low-energy x-rays from the beam because they are of little value for detection but would increase radiation dose to the cargo. The remaining x-rays pass through a collimator, a slab of heavy metal with vertical holes drilled in it, so that only those x-ray photons traveling in the desired direction, downward, can pass through, forming a beam traveling in one direction. Collimation removes from the beam those x-rays traveling in other directions that would interfere with detection by scattering in the cargo or traveling directly from the x-ray generator to the detectors without going through the cargo.

The x-ray beam would pass downward through the cargo, generating other x-rays as described above, some of which scatter backwards. Very few other photons do so. Sodium iodide detectors are placed so as to detect these backscattered x-rays. The detectors are collimated so their field of view intersects with an x-ray beam.[83] **Figure 14** illustrates this configuration. The intersection of

---

[81] The EZ-3D unit can operate without a radiography system. However, the AS&E CAARS system would use an EZ-3D system and a radiography system together.

[82] A Rhodotron is a circular electron accelerator manufactured by Ion Beam Applications. Unlike most linear accelerators, it generates electron beams in continuous waves rather than in pulses.

[83] Sodium iodide detectors are used instead of detectors with a higher resolution, such as high-purity germanium. Since the EZ-3D system requires only that the detectors count photons, as opposed to identifying isotopes by their gamma-ray spectra, detectors with medium energy resolution suffice, and are much less costly than germanium detectors. Since the Rhodotron generates an electron beam in a continuous wave, and thus generates an x-ray beam in a continuous wave (continued...)

a beam and a detector's field of view forms a voxel. This intersection of two lines locates the voxel in two dimensions; the position of the "slice" of the container being examined locates the voxel in the third dimension. Each detector records the number of individual x-ray photons it detects in each voxel. AS&E expects the system to be able to scan a standard 40-foot cargo container for high-Z material in 15 seconds.

Since the backscattered photons would pass through other cargo between the x-ray generator and a voxel, and between a voxel and a detector, the system is designed to account for, and subtract, the effects of this other cargo. Scanning a container produces radiography and backscatter data on how each voxel attenuates x-rays. An algorithm would integrate and analyze both types of data. In effect, to reconstruct the contents of a container, it would create a hypothesis about what is in the container where, and would calculate how closely the hypothesis matched the data. The algorithm would then alter the hypothesis iteratively until it provided a best fit with the data. The system is designed to alarm automatically when it detects voxels containing high-Z material and meeting other conditions. For example, the algorithm might alarm only if it detected a number of contiguous voxels of high-Z material so that it would not alarm each time it detected, say, a lead sinker. The process would repeat for each slice of the container. Korbly states that the algorithm has been shown to work in laboratory demonstrations.

Figure 14. Schematic Diagram of EZ-3D Technique

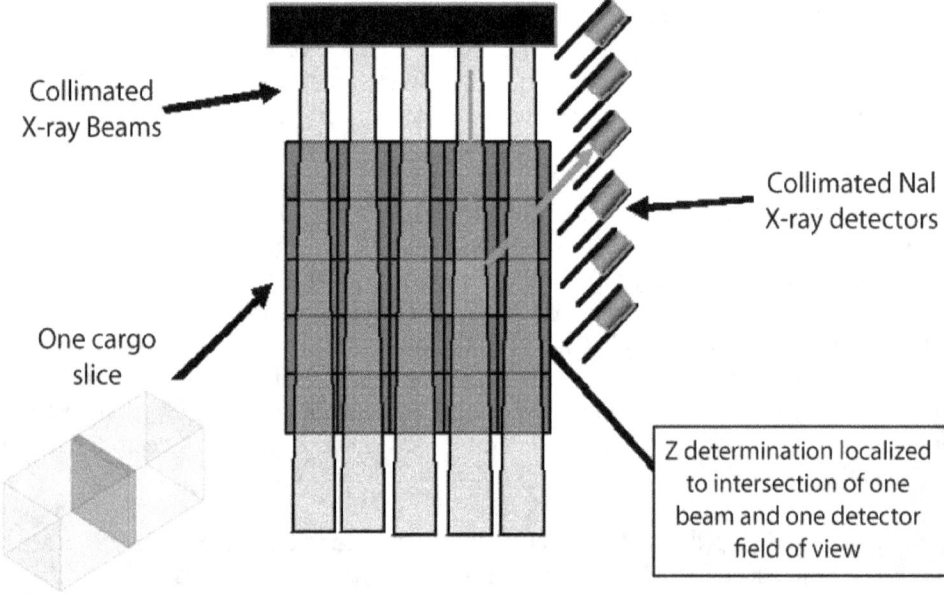

Source: Passport Systems, Inc.

Notes: This figure shows EZ-3D scanning one "slice" of a cargo container. The small diagram at the left depicts a slice, a transverse segment of a container. Many collimated x-ray beams exit from the collimator (bar at top of large diagram). They travel downward through the container. X-ray detectors are set to the side of the

---

(...continued)

with no large spikes in the number of x-rays, the sodium iodide detector is able to record all the time. In contrast, a beam that turned on and off several thousand times a second would produce large spikes in the number of photons, overloading the detector.

container, and collimated so that they view x-rays scattering at a backwards angle (i.e., greater than 90 degrees) from the direction of the beam. The intersection of an x-ray beam and the field of view of a detector forms a voxel whose effective Z is analyzed as shown in **Figure 13**. A beam and a field of view constitute two lines; their intersection locates the voxel in two dimensions. The position of the slice of cargo locates the voxel in the third dimension.

Potential benefits

(1) Jeffrey Illig, the CAARS project manager at AS&E, stated in October 2008 that the AS&E CAARS would, if successful, look at 3-D voxels, rather than 2-D pixels; the latter approach, in effect, provides the average Z of an x-ray traveling through the entire width of a container as in the L-3 and SAIC CAARS systems. As a result, he says, the AS&E system is expected to generate more data by breaking the volume being inspected into smaller units, simplifying the task of the detection algorithm. (2) Joel Rynes, Program Manager of DNDO's CAARS program, said that the EZ-3D effect is more specific to high-Z materials than is dual-energy radiography, potentially improving performance, though it will not be able to discriminate between (for example) lead and uranium within the scan time and dose to cargo limits required.[84] (3) The system is designed so that, if its radiography unit detects an area of dense material, it could direct the EZ-3D x-ray beam to spend more time scanning that area, further reducing the probability of a false positive or false negative. (4) Illig says that the system is designed to specify the location of a suspicious object in three dimensions; as a result, CBP personnel would know where to look in a secondary inspection, easing their task and reducing the time that a container is delayed for inspection. (5) If the system works as anticipated, it is claimed, it would automatically alarm on high-Z material, making it easy to use. (6) Because the x-ray beam is aimed downward, much of it would be absorbed by the ground, reducing the amount of x-radiation that escapes and reducing shielding requirements.

Status, schedule, and funding

AS&E has been working on its CAARS system since late 2006 under a contract that DNDO awarded for system development in September 2006. (The L-3 CAARS section provides the total cost of DNDO's three CAARS projects.) In early 2008, DNDO changed the goal of the CAARS program from system acquisition to an advanced technology demonstration (ATD). Passport Systems has conducted numerous laboratory experiments at the University of California, Santa Barbara, to develop EZ-3D using the university's 5.3-MeV accelerator to generate x-rays to scan a 4 ft x 4 ft x 4 ft volume containing objects used to simulate commercial cargo. DNDO conducted a Critical Design Review (CDR) of the AS&E CAARS in October 2008, i.e., a review of the specifications for all components of the system and the links between them. DNDO approved the system's design at that time. As a result, the design is locked in and AS&E is able to begin to order hardware to assemble an ATD prototype system. Illig stated in early October that the system's design was complete and that, if the CDR is successful, AS&E would begin assembling a full-scale ATD system in November 2008. AS&E would also construct a shielded-room facility to develop the system in its anticipated production configuration. AS&E anticipates obtaining the first data from this ATD system in April 2009. After several months of developmental testing, AS&E would turn the prototype system over to DNDO so that agency could characterize and evaluate the system's performance using its own containers, cargo, and targets. If that phase is completed successfully, DNDO would decide whether to test the prototype

---

[84] Personal communication, October 27, 2008.

under operational conditions, such as having CBP personnel operate it at a port of entry. Successful completion of that phase, in turn, would lead to a decision by DNDO on whether or not to purchase and deploy the system. Illig stated that AS&E could deliver the first production unit around the end of CY2010, with full-scale production commencing in CY2012; it projects the cost of its CAARS system at $8 million to $10 million per unit, assuming a buy of 25 units.[85] However, Rynes states:

> DNDO has no plans at present to purchase and deploy the CAARS systems. We will use prototypes of the three CAARS systems to collect data to feed a cost-benefit analysis that could lead to future procurements by DHS. The AS&E CAARS prototype will be a full-scale laboratory prototype that would still take substantial work to get it ready for a port environment. It will take less work to get the SAIC and L-3 prototypes ready for a port deployment. The AS&E system has always been the high risk, high benefit solution.[86]

Rynes stated that as of February 2009, development of the AS&E CAARS system was on hold due to technical issues that AS&E is trying to resolve.[87]

This report discusses this cost-benefit analysis under "status, schedule, and funding" in the L-3 CAARS section.

Risks and concerns

Scientific risks and concerns

(1) In theory, the system could differentiate between different high-Z materials, such as uranium vs. lead, if the materials were pure chemical elements and certain other conditions were ideal. In practice, so doing would take longer than differentiating between high- and medium-Z material, and impurities and interference from other cargo could make differentiation between different high-Z materials very difficult at best, so any high-Z material in a container could require a secondary inspection, delaying that container. (2) A container can include many types of cargo, and there is no requirement to declare the arrangement of the cargo. It would appear very difficult to develop an algorithm that can reliably eliminate x-rays scattered by cargo between the voxel being interrogated and a detector and reconstruct the locations of different chemical elements within a cargo container.

Engineering risks and concerns

(1) The system has not been demonstrated in a full-up configuration. Passport Systems, a Massachusetts company, uses an accelerator in California for its experiments, and these experiments use a lower-powered accelerator (5.3 MeV) than the Rhodotron (9 MeV). What risks, if any, are associated with shifting from a lower- to a higher-powered accelerator? (2) The integration of a linear accelerator for radiography and a Rhodotron for EZ-3D has not been demonstrated. (3) Passport Systems has simulated the performance of EZ-3D by gathering radiography data separately, bringing these data to the EZ-3D experimental facility in California,

---

[85] Personal communication, October 8, 2008.

[86] Personal communication, October 22, 2008.

[87] Personal communication, February 9, 2009.

merging ("injecting") the radiography data into the EZ-3D data, and feeding the merged data into the cargo reconstruction algorithm. As with any simulation, one might ask how accurately the simulation reflects reality. (4) Will the AS&E system be able to meet the intended scan speed? There is always a risk of problems of this sort when one scales up from a laboratory demonstration system to an operational system. (5) Rhodotrons take a long time to build, and very few have been built. For example, AS&E has the world's 18th unit. Could the manufacturer, Ion Beam Applications (IBA), a Belgian company, build them faster and cheaper? IBA has told AS&E that if IBA gets a large order, it would build a facility in the United States to assemble Rhodotrons. Could it ramp up production in a new facility without major hurdles?

Cost and schedule risks and concerns

(1) If AS&E received a contract for multiple (e.g., 25) production units, could IBA ramp up its Rhodotron production facility on schedule? (2) The AS&E system is expected to be costly, and a main component of cost is the Rhodotron, at €2 million to €5.7 million apiece for single units.[88] Will IBA be able to reduce Rhodotron cost through research and through quantity production?

Operational risks and concerns

(1) Illig states that AS&E has extensive experience in designing systems and their user interfaces for CBP and other front-line users, but that AS&E has not consulted with CBP on the design of its CAARS system. Is that cause for concern, or is the user interface fairly standard by now so that extensive consultation is not needed prior to operational testing? (2) Illig states that the footprint of the unit is 60 ft by 160 ft mainly because it uses concrete walls for radiation shielding. The footprint is a concern for port operators because space is at a premium at ports; it is less of a concern at land border crossings. Illig states that the length could be reduced to perhaps 40-50 ft by 120 ft by using some means other than a truck to pull containers through the system; is that reduction sufficient? (3) The system generates radiation in two ways. A radiography unit uses a 6-MeV linear accelerator, aimed horizontally across a cargo container, to generate x-rays (through the *bremsstrahlung* process). Illig states that there is a "beam dump" to absorb x-rays on the other side of the container, and that the accelerator is pulsed, so that it is off most of the time, reducing the shielding needed. However, some x-rays scatter off detectors and cargo. The Rhodotron is on all the time, and will generate 9-MeV electrons, but is aimed downward so that the Earth absorbs most of the resulting x-rays. Further, the *bremsstrahlung* process generates x-rays in all directions; will the shielding be adequate? Accordingly, CBP and port personnel will be concerned about the amount of radiation that escapes. Will AS&E be able to provide satisfactory assurances on this point?

Potential gains by increased funding

Illig states that added funding would allow AS&E to build a test cell that had a Rhodotron and a linear accelerator for radiography together so that it could obtain actual data. That, he says, would permit engineers to develop algorithms to integrate both types of data using actual data. Added funds, he said, would also let AS&E purchase more test articles, conduct tests on more cargo

---

[88] Information provided by Ion Beam Applications, October 24, 2008. These figures are for a Rhodotron for use in x-ray mode, i.e., with a target for the electron beam to strike to create *bremsstrahlung* x-rays. As of October 30, 2008, €1 = $1.2923.

configurations, and increase test time, all of which would characterize system performance better and reduce risk.

Potential synergisms and related applications

According to Illig, as part of the ATD for CAARS, AS&E would build a facility integrating backscatter and radiography technologies. This facility could provide a testbed for Passport Systems' nuclear resonance fluorescence technology and, more generally, for research into resolution of alarms caused by possible shielded nuclear material. Similarly, Rynes believes that the facility could have various applications:

> The AS&E CAARS prototype is being installed at MIT's Bates Linear Accelerator Center. The AS&E x-ray source (9 MeV, continuous wave) provides a capability that does not exist in the United States. After the CAARS program is complete, DNDO must dispose of all equipment procured under the contract. One option proposed is to keep the source at the Bates Center and establish the source as a "user facility" where researchers can come in to do experiments (e.g., NRF measurements at 9 MeV or experiments with prompt photofission). Another option is to let AS&E keep the source to use in a possible follow-on system (e.g., pilot deployment of a modified CAARS design). It is too soon to make this decision; it depends on how well the AS&E CAARS system performs in its upcoming tests.[89]

Muon Tomography[90]

The problem

Nuclear weapons or SNM may be hidden in cargo containers, automobiles, or elsewhere. Radiography might detect a fully assembled nuclear weapon but might miss a small piece of HEU, depending on its size and shape, and passive radiation detection might or might not detect lightly shielded HEU, depending on such factors as the amount of HEU, the thickness and type of shielding, and whether it contained trace amounts of uranium-232, which has a high-energy (2.614-MeV) gamma ray from thallium-208 associated with its decay.[91] Beams of neutrons or high-energy x-rays or gamma rays hold the potential to detect SNM by itself or in complete weapons by inducing fission. But the radiation emitted by such beams could require shielding and some standoff distance, possibly making it impractical where space is at a premium. Other properties of SNM are its high density and high Z. Such materials cause much greater deflection of muons, a naturally occurring subatomic particle, than do lower-density, lower-Z materials. Detection using muon tomography (MT) would not use a radiation source, avoiding concerns about radiation exposure or salvage fuzing. Yet while MT has been demonstrated in the

---

[89] Personal communication, October 22, 2008. The home page for the Bates Linear Accelerator Center is http://mitbates.lns mit.edu/bates/control/main.

[90] Michael Sossong, Director of Nuclear Research, Decision Sciences Corporation, and Guest Scientist, Los Alamos National Laboratory, and Mell Stephenson, Executive Director of Government Programs, Decision Sciences Corporation, provided detailed information for this section, personal communications, April 2008-February 2009. Others have commented as well to provide alternative perspectives.

[91] A lump of plutonium, whether shielded or not, seems an implausible threat because it would be very difficult for terrorists, by themselves, to fabricate a bomb using plutonium. It would be even harder for them to fabricate such a bomb inside the United States using plutonium they had smuggled in because they would need to take added measures to avoid detection.

laboratory, it remains to be seen if it can be converted to a system that will work in the real world. At issue: Is muon tomography an operationally feasible means of detecting nuclear weapons or SNM in the presence of clutter from actual cargo? Is it a cost-effective means of detecting high-Z material in vehicles with people inside them, and in cargo at ports of entry and choke points, without affecting the flow of commerce?

Background

Muons are heavy subatomic particles generated when cosmic rays strike atoms in the Earth's upper atmosphere. Most muons travel at over 95% of the speed of light.[92] Given their speed and mass, they are highly energetic, with a mean energy of 3 billion electron volts.[93] As such, they are highly penetrating. For example, they can penetrate 1.3 m of lead or 15 m of water.[94] About 1 muon strikes each square centimeter of the Earth's surface per minute.[95] Matter deflects muons, with the degree of deflection determined statistically by the density and Z of the matter. As it happens, there is large separation between the average angle of deflection resulting from low-Z, low-density material in commerce, like food or plastic, and medium-density, medium-Z material like steel, and between the latter and high-Z, high-density material like tungsten, lead, or SNM.[96]

Tomography divides a solid object into many parts, determines a characteristic (e.g., density) of each part, and assembles the parts into an image of the object. A medical CAT (computed axial tomography) scan, for example, creates images of each slice (about 1 to 10 mm wide) of the body part being scanned. Muon tomography measures the trajectory of each individual muon before it enters a cargo container and again after it exits. In simplest terms, the intersection of the trajectories indicates the angle of deflection (and thus high, medium, or low density and Z) and the point of deflection, though the trajectory is more complex because a muon interacts with many atoms as it passes through a container. An algorithm integrates data from numerous muon trajectories to form a three-dimensional image of the container based on the density and Z of its contents. The statistical difference in deflection between high-Z, high-density material and other material, combined with muons' high penetrating power, is the basis for an MT system to detect SNM, whether in weapons or by itself.

Technology description

Decision Sciences International Corporation (DSIC) is developing an MT system for use with vehicles and containers. (DSIC changed its name from Decision Sciences Corporation in August 2009.) Development began through a Cooperative Research and Development Agreement (CRADA) with Los Alamos National Laboratory (LANL). **Figure 15** is a schematic drawing of the system. The prototype works as follows. To determine the track of muons, it uses "drift tubes," which are similar in shape to fluorescent light tubes. They are made of aluminum, with a wire running lengthwise through the center of each tube, and are filled with a mixture of gases

---

[92] Personal communication from Michael Sossong, April 23, 2008.

[93] Jonathan Katz, Karol Lang, and Roy Schwitters, "Muon Tomography—The Future of Vehicle and Cargo Inspection," report prepared for Decision Sciences Corporation, July 19, 2007, p. 5.

[94] Ibid.

[95] Ibid., p. 4.

[96] Ibid., pp. 9-10.

typically used in drift tubes.[97] The tubes are arranged in arrays. Each array consists of 12 cross-hatched layers of tubes, alternating between 2 layers in the "x" direction and 2 in the "y" direction. A positive charge is applied to the wires. A muon that strikes the gas in a drift tube creates a trail of free electrons, which are drawn ("drift") to the wire by the positive charge at a known speed. These tubes measure the distance between the wire and a muon's closest approach to the wire. Two layers of "x" tubes establish a "y" measurement, and two layers of "y" tubes establish an "x" measurement. Combining data from each set of 4 tubes establishes a point on a muon's trajectory, and the system uses 3 points to define the trajectory.

DSIC had originally planned to have the tubes in a "tunnel" configuration, as illustrated on the left side of **Figure 15**, with arrays of drift tubes on the top, bottom, and sides of the object being examined to determine the incoming and exiting trajectories of individual muons, but has instead decided to use a "top/bottom" configuration, as shown on the right side of the figure, with arrays on top and bottom only. This configuration would be less costly than the tunnel, but it would require up to 10% more scan time because muons entering from or exiting to the side would not be counted. That increase could be reduced by using multiple units arrayed side by side to scan multiple lanes of traffic so as to record the tracks of more muons entering and exiting the object being examined. As with any detection system, total throughput (e.g., number of tractor-trailer trucks exiting a port per hour) could be increased by deploying more units. Also in this configuration, units are only as wide as a lane of traffic, unlike many other systems. DSIC views this as important because space is severely limited at many ports and, even where it is not, it could be difficult to rearrange traffic lanes to accommodate wider detection equipment.

Figure 15. Schematic Drawing of Muon Tomography Inspection Station Configurations: Tunnel and Top/Bottom

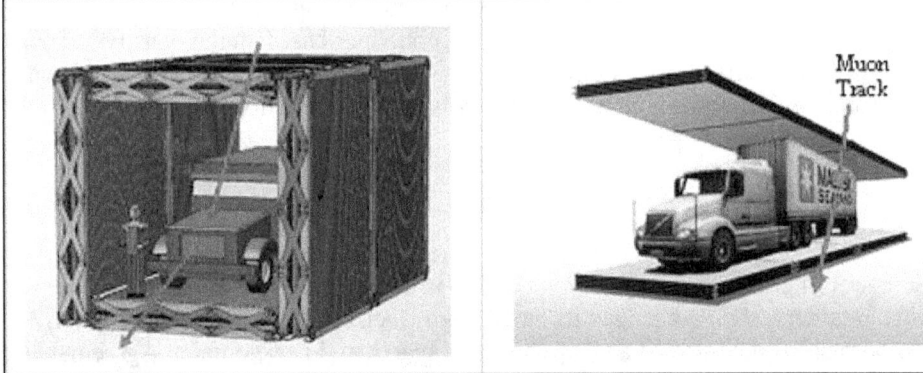

Source: Decision Sciences International Corporation.

Notes: The station is designed to detect the scattering angle of individual muons as they enter and exit the truck and its cargo. DSIC initially researched the tunnel configuration (left); it currently focuses its efforts on the top/bottom configuration (right).

For MT to work, the system must identify each muon uniquely so it can match entry and exit tracks. According to DSIC, the system's electronics can do this because muons travel at essentially the speed of light, and entry and exit tracks occurring at the same time are easily

---

[97] The mixture currently includes a small fraction of helium-3, which is of use for detecting neutrons. Recognizing the scarcity of that gas, DSIC would replace helim-3 with helium-4 (the gas used to fill balloons, for example) to maintain the same proportion of gases, and would use boron-10 to detect neutrons.

matched up. The electronics can keep up with the tracks, since the rate at which muons strike the drift tubes is far lower than the rate at which the electronics can process muon signals. Some 3,000 muons strike the drift tube array each second, and the electronics can read a muon hit in a millionth of a second. The odds of two muons striking a drift tube within a millionth of a second are 3000 out of 1,000,000, or 0.3 percent, so MT can uniquely identify a single muon 99.7% of the time.

If no matter at all were present between the top and bottom drift tube arrays, a muon's exit track would be a straight-line continuation of its entry track. If a muon interacted at only one point, the intersection of the two tracks would indicate the angle and point of deflection of an individual muon. The point of deflection would locate a voxel,[98] and the angle of deflection would indicate "scattering density," a combined measure of Z and density of the material in that voxel.[99] In practice, a muon interacts with all matter along its path, displacing the exit track from a straight line. The amount of displacement provides information on the scattering density, location, and thickness of the material.

From these data, an imaging algorithm calculates the degree of scattering of muons for each voxel and creates a 3-D image of the contents of the object being scanned. Resolution of the scan increases with time, as **Figure 16** shows, as each pair of muon tracks adds data. The image is displayed on a computer screen and can be rotated so the viewer can visualize it as if in three dimensions, as **Figure 17** shows. Based on computer simulations and laboratory tests conducted in early 2010 using prototype equipment, DSIC states that MT can differentiate between high-Z and medium-Z material, so it can pick out HEU hidden in a cargo of steel parts, or even hidden as a piston inside an engine.

---

[98] A voxel is a volume element, analogous to a two-dimensional pixel, or picture element.

[99] Deflection is influenced both by Z and density. A muon is more likely to interact with a larger atom (higher Z) than with a smaller one. A muon is also more likely to interact with atoms the closer they are packed together (density). Scattering density combines Z and density into one unit.

Figure 16. Muon Tomography Resolution Increases with Scan Time

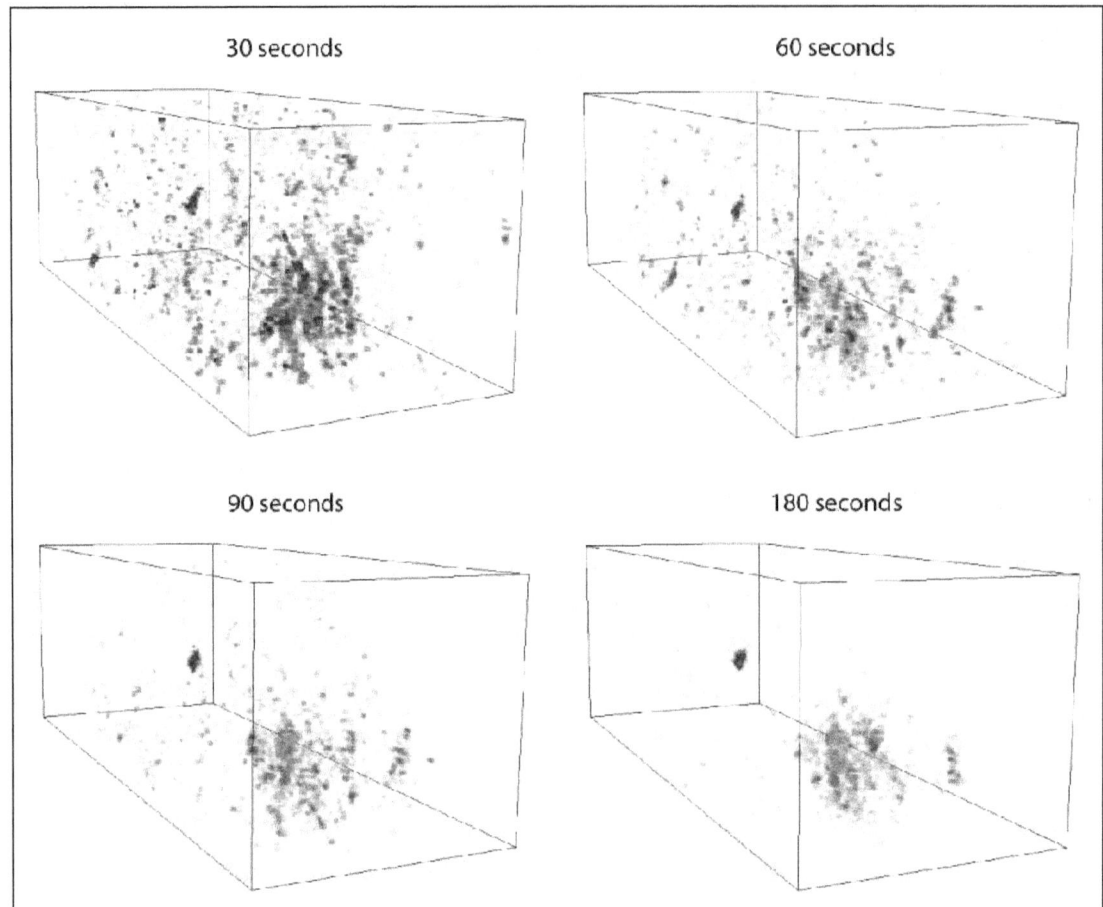

Source: Decision Sciences International Corporation, April 2010

Notes: This figure shows scans of a car at various times using actual data. The dark spot "floating" at the back of the image is high-Z material and is highlighted in red.

**Figure 17. Muon Tomography Creates
Three-Dimensional Images**

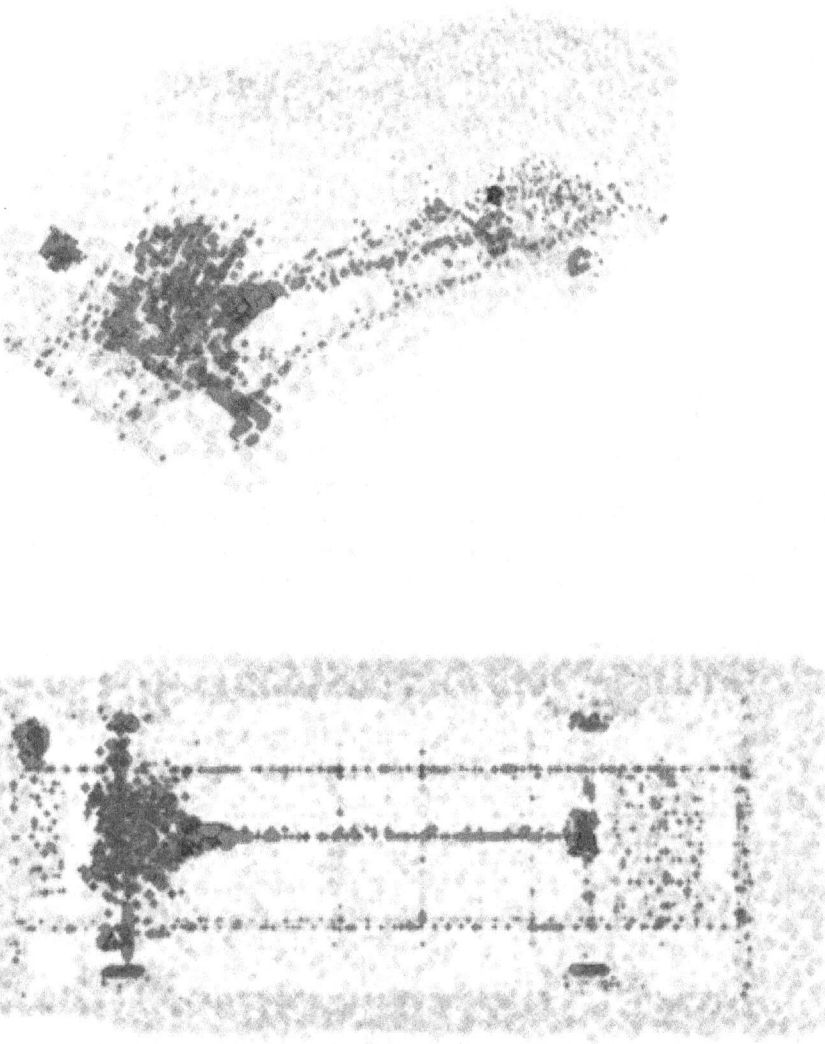

Source: Decision Sciences International Corporation.

Notes: This figure shows one simulated scan at 90 seconds of side-angle and top views using simulated data. The dark spot above the rear axle represents SNM and is highlighted in red.

Based on tests conducted in early 2010 with its prototype scanner, DSIC estimated in April 2010 that its system would take less than 1 minute to clear most containers with 95% confidence; that it could automatically detect a cube of unshielded SNM 5 cm on a side in a container in that time; and that it could automatically detect that cube inside shielding (e.g., a larger cube) in less time. Others disagree, as discussed under "Scientific risks and concerns," below. DNDO observed, "These performance results are not supported by data collected during the DNDO TRR [test readiness review] demonstration conducted January 11-14, 2010."[100] As with any system,

---

[100] Information provided by Leon Feinstein, DNDO, e-mail, May 17, 2010.

increasing scan time would increase confidence while having a greater impact on the flow of commerce.

The DSIC scanner can also detect gamma rays because they produce a signal different from muons when they strike drift tubes. According to DSIC, when a gamma ray strikes an aluminum drift tube, it knocks electrons off the aluminum, which ionize atoms in the gas inside the drift tubes. These electrons drift to the central wire and are recorded. The system can differentiate between electrons generated by gamma rays and by muons, it is claimed, because the muon-tracking algorithm can compute a track for muon-generated electrons but not for gamma-generated electrons, as the latter do not pass through a drift tube. A gamma-ray count above background levels would be suspicious, though there are many innocent gamma-ray emitters in commerce. A low count might reduce suspicions, though vehicles in adjacent lanes could suppress background gamma rays.

The scanner has no spectroscopic capability, so it cannot identify the source of gammas by their spectra. However, DSIC argues that its scanner uses the presence or absence of gammas to reduce scan time. By way of background, HEU that has been through a nuclear reactor picks up a small amount of uranium-232, which has an extremely energetic gamma ray (2.614 MeV) from thallium-208 associated with its decay, making it very hard to shield. Even if HEU has not been through a reactor, such as if it has been produced by centrifuge from natural uranium, it will have some uranium-238, which has a gamma ray of 1.001 MeV. DSIC states:

> The gamma signal from any HEU has a 1.001-MeV component from its uranium-238 content. Even for uranium enriched to 93 percent uranium-235, with 7 percent uranium-238, shielding 1.001-MeV gammas so they fall below the system's detection threshold of about 20,000 gammas per second requires about 1.4cm of lead. This increases the size of the threat object we're searching for with muon tomography to almost 9 cm in diameter, compared with 6 cm for a 2-kg cube of unshielded HEU. Since the number of muons passing through an object is proportional to its area, about 2.25 times more muons will pass through the larger object than the smaller one. This means muon tomography can clear the shielded package 2.25 times faster than the bare HEU. Because only some fraction of containers emits gammas, this allows us to clear the vast majority of containers for the large threat package in 1/2.25 the time (26 seconds). If we were to search for the bare HEU every time, the average time to clear would be around 60 seconds. This makes a huge difference in our throughput rates. Containers with cluttered scenes or potential shielding objects would require longer scan times, but these scenes should be rare. Throughput is determined by the average scan time, so averaging a majority of 26-second scans with a few 60- or 90-second scans would have little effect on throughput. We don't know the number of occurrences in commerce of these types of cluttered scenes, so our next step is to scan containers in the flow of actual commerce and adjusting our scan procedures according to what we find.[101]

DSIC also states that, because its system has several thousand drift tubes, which function as detector elements, it can provide a general location of gamma-ray emitters in a container, helping to distinguish between point sources like a small amount of plutonium and distributed sources like kitty litter. It further claims that this capability can help clear false positives and determine which parts of a container warrant closer examination by the system's muon tomography element.[102]

---

[101] Information provided by Michael Sossong, DSIC, e-mail, May 4, 2010.

[102] Information provided by Michael Sossong, DSIC, e-mail, May 9, 2010.

DNDO finds the argument that gamma-ray detection capability adds to the efficacy of muon tomography to be unconvincing.[103] "The attenuation of 1-MeV gamma rays by 1.4 cm of lead can also be achieved by about 2 cm of iron or 6 cm of aluminum or 16 cm of water; in other words, the HEU sample can be easily shielded by innocuous materials commonly found in cargo and effectively invisible to the MT system." As a result, DNDO argues, gamma-ray detection capability would probably not reduce MT scan time in actual cargo, much less in a container in which terrorists had arranged the cargo so as to reduce the probability of detecting a nuclear bomb. DNDO prefers that DSIC would concentrate on proving the concept of MT before trying to augment it with detection of other types of radiation.

Based on testing, improvements in electronics to reduce noise, and computer modeling, DSIC anticipated in April 2010 that MT would be able to distinguish between SNM and medium-Z materials in under a minute. Also in that month, DSIC estimated that perhaps 2 to 5 minutes would be needed to distinguish between HEU and lead and about twice that to distinguish HEU from tungsten.[104] These figures are about half those of estimates made in July 2009. An extended scan of a container would be easier, less costly, less intrusive, and faster than unloading a container, inspecting its contents, and reloading it, a process that could easily take hours.

Christopher Morris, a physicist at Los Alamos who has done extensive research on MT, stated,

> If muons were to identify a shielded container, I would advise taking as long as reasonable (perhaps ~1 hour) to survey a container with muon tomography before taking the risk of any invasive action. In this longer scanning time, one should be able to provide detailed images of the configuration of a threat object, estimate its yield if it is in a weapon configuration, distinguish between different high-Z materials radiographically, and carefully study the passive radiation signatures. This might avoid triggering a salvage-fuzed weapon.[105]

Detection might be made faster and more accurate by an algorithm that would subtract the known part of an image, making it easier to focus on suspicious objects. A library would be constructed with MT images of many models of trucks, cars, containers, and trailers. When a truck entered the MT system, a scanner would read its license plate. The algorithm would call up the MT image of that truck from the library, and would subtract that image, voxel by voxel, from the image of the truck generated by the MT system, leaving an image only of the cargo and any anomalous objects in the truck itself. According to DSIC, scans performed on several vehicles have demonstrated the efficacy of this approach. Some, though, question whether the algorithm could handle correctly any variant to the vehicle (e.g., a modification to the engine). As of May 2010, DSIC had constructed models of several vehicles and was evaluating their utility. The library of vehicles will grow as scans of various vehicle models are performed. This library is not a critical piece of the technology for scanning cargo containers, but would be of more value for scanning passenger cars.

To improve its ability to detect SNM, DSIC had initially designed a prototype scanner with the ability to detect neutrons as well as muons and gamma rays. However, DSIC felt that the ability to track muons and to detect and locate gamma ray emitters sufficed As of April 2010, it had postponed plans to use its system to detect neutrons while keeping it as an option. The optional

---

[103] Information in this paragraph provided by Leon Feinstein, DNDO, e-mails, May 17 and 19, 2010.

[104] Information provided by DSIC, e-mail, April 28, 2010.

[105] Personal communication, July 15, 2008.

neutron detection capability would use drift tube walls coated with boron-10.[106] Absence of neutrons may help clear false positives, though hydrogenous cargo could absorb neutrons, preventing them from being detected. There are few sources of neutrons in commerce;[107] accordingly, the presence of neutrons would be suspicious.[108]

Potential benefits

(1) Because it uses naturally occurring muons, muon tomography does not use a source to generate radiation, so it does not require shielding or a standoff distance to protect workers from radiation. (2) Since it does not generate radiation, it cannot harm people or damage the contents of containers, so it could be used at land border crossings to search cars for SNM while the passengers are inside, speeding the flow of traffic. (3) Because it does not generate radiation, it cannot trigger salvage-fuzed weapons. (4) DSIC claims that MT is very unlikely to show a mass of high-Z material where none exists, though DNDO stated that DSIC did not demonstrate this at the TRR. Reducing the false alarm rate is important to CBP because each alarm, whether true or false, may require considerable effort to clear. (5) Because muons are highly penetrating, they can be used to detect high-Z material even when shielded. (6) The detection algorithm is intended to detect, locate, and alarm for high-Z material automatically, greatly reducing the need for human interpretation. The display algorithm shows the operator where high-Z material is located and may provide information about its shape. (7) The top/bottom configuration is narrow enough to scan trucks leaving seaports within existing lanes, avoiding the need to modify traffic patterns.

Status, schedule, and funding

In 1995-1996, LANL built and tested a small MT prototype (6 ft by 6 ft scan area) to demonstrate the detection of high-Z materials. Based on simulations and test data, DSIC and LANL signed a CRADA in May 2007 to commercialize LANL's MT technology.[109] Since then, DSIC has provided about $7 million to LANL as part of the agreement. Under this funding, LANL and DSIC staff built a larger scanner, 12 ft high by 12 ft wide by 16 ft long, large enough to scan a portion of an SUV, that was operated at LANL from October 2008 to mid-June 2009 for a cost of $1 million. Vehicles were tested in the device with various clutters as well as medium- and high-Z materials. DSIC moved the scanner to its headquarters in Poway, CA, and modified it to a top-

---

[106] According to DSC, "Thermal [low-energy] neutrons undergo reaction with the boron-10 nuclei, forming a compound nucleus (excited boron-11) which then promptly disintegrates to lithium-7 and an alpha particle. Both the alpha particle and the lithium ion produce closely spaced ionizations in the tube gas, permitting the system to count neutrons." Personal communication, Michael Sossong, June 30, 2009. The initial plan was to use helium-3, but as of April 2010 the plan calls for the scanner to use boron-10. Because of its scarcity, helium-3 may well be unavailable for neutron detection; see Steve Fetter, "Overview of Helium-3 Supply and Demand," presentation at American Association for the Advancement of Science workshop on helium-3, April 6, 2010, http://cstsp.aaas.org/files/he3_fetter.pdf.

[107] Richard Kouzes et al., "3He Alternatives for National Security," presentation to American Association for the Advancement of Science workshop on helium-3, April 6, 2010, slide 6, http://cstsp.aaas.org/files/he3_kouzes.pdf.

[108] According to DSC, "In another SNM detection technique, the time correlation of a muon stopping in the volume can be correlated with a burst of neutrons from muon-induced-fission in the SNM. This would provide a positive signal of the presence of SNM and when fused with other signals from the system, could provide faster, more accurate scanning." Personal communication, Michael Sossong, June 30, 2009.

[109] Decision Sciences Corporation, "Decision Sciences Corporation Announces Agreement with Los Alamos National Laboratory to Collaborate on Homeland Security," press release, May 3, 2007, p. 1.

---

bottom configuration 12 ft high by 16 ft wide by 24 ft long, large enough to scan SUVs, automobiles, and 20-foot shipping containers.

DSIC canceled a Test Readiness Review (TRR) scheduled for April 2009 because of problems with the muon tracking algorithm and because the decision algorithms needed further development and testing.[110] As of July 2009, DSIC planned to conduct a TRR in late September or early October 2009. A TRR shows whether the system is ready to begin a Proof-of-Concept (PoC) demonstration, the last phase of Exploratory Research at DNDO. DNDO conducted a TRR in January 2010 at DSIC's facility in Poway, CA. R. Leon Feinstein, DNDO program manager for near-term testing of the DSIC MT prototype, described the results as follows:

> The goal of the TRR was to determine if the MT prototype was ready for a DNDO-sponsored Proof-of-Concept (PoC) demonstration that would more fully characterize and evaluate MT technology. Following the TRR, DNDO's test team concluded that the DSIC MT system was not ready for the PoC demonstration. The team recommended that DSIC continue with its planned hardware and decision algorithm upgrades, addressing issues identified by the test team that limit the functionality and performance of the MT prototype. DSIC had already recognized these needed upgrades and fixes at the time of the TRR, and planned to complete them by the end of May 2010.
>
> DNDO will conduct a second TRR after the following conditions have been met: (1) DSIC must complete its proposed upgrades and fixes; and (2) the DSIC TRR Report (a contract deliverable) must be revised, re-submitted and approved by DNDO.
>
> DNDO rejected DSIC's first draft TRR report of February 8, 2010, for the following reasons. (1) It described demonstrations and measurements that the DNDO test team neither observed nor recorded. (2) It focused on gamma-ray detection results that were not part of the TRR and were not observed or recorded by the DNDO team. (3) It did not address numerous anomalies and instabilities in the image reconstruction and decision algorithm for detection of high-Z, high-density metals. These anomalies and instabilities are quite serious and were carefully recorded by the test team during the TRR and briefed to DSIC at the TRR conclusion. DSIC did not address these issues in its first test report, such as by providing a technical explanation and possible mitigation strategies. (4) Many comments and conclusions in the report are inconsistent with observations by the test team and are not backed up with empirical data or technical analysis.[111]

Risks and concerns

Scientific risks and concerns

A concern is that terrorists might be able to counter MT. For example, they might try to smuggle uranium through a detector in pieces below the system's detection threshold, but that would risk exposing the plot to detection many times. Alternatively, since MT indicates the Z and density of a voxel, it might be possible to reduce those characteristics of HEU by forming it into pellets and mixing them with a low-Z substance. These techniques, however, would require fabricating HEU into a weapon-usable shape within the United States, introducing considerable difficulty and

[110] Information provided by R. Leon Feinstein, Transformational and Applied Research Directorate, DNDO, program manager for near-term testing of the DSIC muon tomography prototype, personal communication, July 29, 2009.

[111] Information provided by R. Leon Feinstein, personal communication, May 10, 2010.

providing clues to the plot. U.S. nuclear weapons use hollow pits,[112] in which a hollow shell of SNM is imploded to form a supercritical mass, initiating a nuclear explosion. Other nations might use this method as well. If terrorists were to obtain a state-made hollow-pit nuclear weapon, MT might not detect a pit of this type if the SNM shell were thin enough and voxels large enough that SNM did not fill most of a voxel.

Also at issue is the amount of time required for a scan. The Royal Society, the U.K. national academy of science, issued a report in March 2008 based on a meeting with dozens of experts from many nations, and stated,

> The key limiting factor is the time required for muon radiography, up to several hours to image only a cubic foot of a block of iron. According to a detector concept being developed by Los Alamos National Laboratory (LANL), it would take four minutes to image a cargo container. However, this would require detector panels perhaps the size of a large room. Moreover, once a shielded source has been identified it may take several hours to unpack the cargo to locate it.[113]

DSIC responds that since these concerns were raised, it has made significant advances in extraction of signal from muon scattering data, such as using 3-D images (tomography) rather than 2-D images (radiography). It also claims that increasing scan time from 1 to 5-10 minutes may permit differentiation between SNM and other high-Z material, avoiding the need to unpack a cargo container.[114] According to another analysis, "only a few scattered muons are required to determine Z accurately enough to distinguish among the major groups of Z [high, medium, and low] with high confidence, and the value of Z is conveniently displayed as a color image."[115] In contrast, Feinstein said in May 2010,

> It takes many muons to distinguish *with high confidence* high concentrations of high-Z nuclei found in actinide metals [e.g., uranium and plutonium] from, say, iron and lead in a cargo filled with medium-Z clutter, particularly in a vertical direction. A 10x10x10 [cubic centimeter] voxel at sea level would be penetrated by about 200 muons on the average in 2 minutes of interrogation. The minimal amount of time to discriminate a liter of SNM from a liter of lead has not yet been determined or demonstrated in a densely cluttered cargo environment. MT estimates the nuclear charge concentration in each reconstructed image voxel and, hence, cannot detect high-Z material that is significantly diluted by air or other low density, low to medium Z material in the same voxel; and will not have sufficient sensitivity to discriminate SNM from DU [depleted uranium, i.e., uranium with most U-235 removed] and other relatively high-Z material.[116]

Robert Mayo, Program Manager, SNM Movement Detection/Radiation Sensors, and Advanced Materials Programs, Office of Nonproliferation R&D, NNSA, raised other concerns in 2008:

---

[112] U.S. Department of Energy. Office of Declassification. "Drawing Back the Curtain of Secrecy: Restricted Data Declassification Policy, 1946 to the Present, RDD-1." June 1, 1994. Item V (C) (2) (s), at https://www.osti.gov/opennet/forms.jsp?formurl=document/rdd-1/drwcrtf3 html#ZZ1.

[113] The Royal Society, "Detecting Nuclear and Radiological Materials," RS policy document 07/08, March 2008, p. 6, at http://royalsociety.org/displaypagedoc.asp?id=29187.

[114] Personal communication, DSIC, e-mail, April 28, 2010.

[115] Katz et al., "Muon Tomography—The Future of Vehicle and Cargo Inspection," pp. 19-20. DSIC stated in April 2010 that the number of muons required for this purpose is 31; personal communication, April 28, 2010.

[116] Information provided by Dr. R. Leon Feinstein, Transformational and Applied Research Directorate, DNDO, program manager for near-term testing of the DSIC muon tomography prototype, e-mail, May 18, 2010.

While muon tomography has been demonstrated to locate regions of interest in cargo that contains dense material, its material identification abilities are limited. What's more, quite large sized detection equipment, much more so than in other detection systems, is required, making MT rather impractical for many nonproliferation and national security applications. Most critically, however, MT is likely to require rather long scan times to adequately resolve dense images in cargo with a reasonable rate of false alarms. For these reasons, it is considered impractical as a screening technique. There are much more practical threat material identification and characterization tools being developed by various agencies of the federal government including passive spectroscopic and active detection technologies, as well as advanced radiography, all of which could be advanced and operationalized much more quickly with increased support.[117]

DSIC claimed in May 2010 that data from actual tests (as distinct from modeled data) do not support the concerns raised in the preceding two paragraphs.[118]

Engineering risks and concerns

The MT system as it stood in April 2010 will incorporate gamma-ray detection (though not identification) as well as muon detection, and retains the option to incorporate neutron detection as well. This plan raises several questions.

First, how valuable is the gamma-ray detection capability for reducing scan time? DSIC's plan is that if the scanner detects gamma rays, it would look for an unshielded SNM source rather than a shielded one. Since the latter would be larger than the former because of shielding, and since MT could find a large object more easily than a small one, the argument goes, this procedure would reduce scan time. But there are many sources of gamma rays in cargo.[119] There is also background gamma radiation from such sources as uranium. What fraction of cargo containers in actual commerce contain gammas? DSIC plans to conduct scans on containers in the flow of commerce to help answer this question.

Second, what confidence could there be that absence of a gamma-ray signal would permit a reduced scan time? DSIC suggests that scan time could be reduced to 26 seconds. Yet as **Figure 16** shows, a 30-second scan of a car contains a considerable amount of clutter. Clearing cluttered scenes in 30 seconds would probably not be a concern for a container full of lettuce, paper towels, or water bottles, but could be a problem for a container with dense metal objects like car parts.

Third, helium-3 is the "gold standard" of neutron detection, but given the shortage of it, DSIC worked with Los Alamos to use boron-10 compounds in the tubes for neutron detection, resulting in a much lower cost to add neutron and gamma capability to the scanner. Could boron-10 be expected to work adequately—not as well as helium-3, but adequately—for neutron detection? If not, might some other combination of gases work adequately?

---

[117] Personal communication, July 31, 2008.

[118] Information provided by DSIC, e-mail, May 21, 2010.

[119] Richard Kouzes, Pacific Northwest National Laboratory, lists the following: "agricultural products like fertilizer, kitty litter, ceramic glazed materials, aircraft parts and counter weights, propane tanks, road salt, welding rods, ore and rock, smoke detectors, camera lenses, televisions, medical radioisotopes" (but not bananas, contrary to public perception). Richard Kouzes et al., "3He Alternatives for National Security," presentation to American Association for the Advancement of Science workshop on helium-3, April 6, 2010, slide 6, http://cstsp.aaas.org/files/he3_kouzes.pdf.

Cost and schedule risks and concerns

Is the schedule too optimistic? In October 2008, DSIC stated that development was proceeding at a rapid pace. It envisioned production after the third-generation prototype was built and field-tested at an operational facility. It stated that the detection algorithms were being modified into a form suitable for operational use. DSIC planned a third-party test validation and life expectancy analyses and experiments. As of October 2008, DSC planned to complete all these steps by the end of 2010. Earlier, DSIC had planned to complete them by the end of 2009. As of July 2009, the schedule had been delayed for several reasons, such as moving the detector to a sea-level location (from Los Alamos, NM, to Poway, CA). As of July 2009, DSIC expected to complete the proof-of-concept demonstration for DNDO in November 2009, to begin construction of a commercial scanner in January 2010, to perform field testing during 2011, and to have the scanner commercially available in 2012.[120] As of April 2010, DSIC stated that that schedule remained current. It had begun design and component testing to increase deployability, manufacturability, and lifetime of its system. The schedule risk is that a system in prototype development must complete many steps before it could be commercially available, yet a problem with any step could delay the schedule. A related concern is whether it is appropriate for DSIC to develop a schedule for commercialization before completing its TRR and PoC demonstration successfully and then moving to an Advanced Technology Demonstration effort. The latter would provide a full characterization of system performance, giving DNDO the data it would need to conduct tradeoff studies for potential applications.

Operational risks and concerns

One operational issue is the clearing of alarms due to innocuous high-Z objects. MT is expected to differentiate between high-Z material and low- or medium-Z material faster than it could differentiate between various types of high-Z material, such as tungsten vs. uranium, though estimates of scan times differ. However, Feinstein stated, "MT is likely a poor choice to discriminate uranium from tungsten. SNAR [shielded nuclear alarm resolution] technologies will likely do this more reliably and much faster. Further, all SNAR signatures have passed their PoC demonstration and evaluation."[121]

If MT proves able to differentiate between SNM and other high-Z material as well as its developers anticipate, then performing an extended-time secondary scan of a container to resolve a high-Z alarm would be faster, simpler, and less costly than unpacking a container. This anticipation, though, is based on experiments using medium-scale laboratory equipment. If experiments find that MT cannot differentiate between SNM and other high-Z material as well as anticipated, then more intrusive means might be required. However, Katz estimates that these alarms would be few and could be cleared easily:

> I think the false positive rate would be a small fraction of a percent. There just aren't that many lead or tungsten ingots in commerce. When they are shipped they sit on pallets on the floor of a container (strapped down) and the rest of the container is empty. 30 tons of lead fills about 2.7 m^3 [cubic meters] out of the 64 m^3 container volume, so it is mostly empty space. A quick look is enough for secondary inspection.[122]

---

[120] Information provided by Michael Sossong, e-mail, July 7, 2009.

[121] Personal communication, May 10, 2010.

[122] Personal communication, July 15, 2008.

While this approach would reduce the burden of clearing high-Z alarms (for other systems as well as MT), the concern regarding nuclear smuggling is not clearing ingots of tungsten or lead shipped in this manner, but detecting SNM hidden in cargo. Further, the false alarm (false positive) rate depends on the threshold that DSIC chooses for detecting small amounts of SNM. Smaller voxels and lower thresholds might be needed to reduce false negatives, but as voxel size and threshold decrease, the false positive rate will inevitably increase, as discussed under "Computer Modeling to Evaluate Detection Capability."

Another issue is the importance of reducing scan time from 60 to 26 seconds. The concept of operations that DSIC envisions as of April 2010 is to use MT with a low-energy (1-MeV) x-ray source for radiography in a primary inspection mode, and to use MT to clear questionable containers in a secondary inspection mode.[123] But CBP agents would probably need more than 26 seconds to examine a radiograph visually for signs of contraband, which in this case would be done in primary inspection. Regarding secondary inspection, a multi-minute scan time is acceptable, as one alternative, unloading a container for inspection, would take much longer, though other alternatives are under development, such as those in DNDO's Shielded Nuclear Alarm Resolution program. A related issue is why DSIC is considering a low-energy x-ray system when higher-energy x-rays, in the range of 6 to 9 MeV, are considerably more penetrating.

A former concern was that many ports could not accommodate an MT system as wide as the original tunnel configuration. In response, DSIC designed a top/bottom scanner the width of existing truck lanes at ports.

Potential gains by increased funding

DSIC states that added funds would enable it to improve the detectors and detection algorithms to provide more detailed imaging; incorporate sensors for additional signatures into the MT system to detect more threats and contraband; expedite the integration of gamma ray, neutron, and muon signals; develop the CONOPS that governs how the system would be used; and enhance and expedite engineering and manufacturing of the production version. Added funds, DSIC says, would also support development of a library of MT images of different vehicle types, as discussed above.

Potential synergisms and other applications

The detection algorithm draws on those used for positron emission tomography, a medical technique to image bodily processes.[124] DSIC is working with the positron emission tomography/single photon emission computed tomography imaging group at the University of California, Davis, to develop advanced imaging techniques. DSIC states, "These algorithms will be applied to detection of SNM in our system and will feed back into the development of algorithms for medical imaging and lesion identification."[125]

---

[123] Personal communication, Michael Sossong, DSIC, e-mail, May 5, 2010.

[124] For further information, see State University of New York at Buffalo, Department of Nuclear Medicine, Center For Positron Emission Tomography, "Positron Emission Tomography," at http://www.nucmed.buffalo.edu/prevweb/petdef.htm.

[125] Personal communication, Michael Sossong, July 12, 2008.

Another role for MT may be as a secondary inspection system for x-ray inspection, such as for containers at seaports. A low-energy x-ray system could quickly clear containers carrying low-density material but not necessarily containers carrying thick, dense cargo. In such cases, MT could interrogate the dense regions for SNM. DSIC claims, "In this role, the scan time requirements imposed on a primary system would be reduced, while still maintaining an effective, high-throughput primary scan and a secondary [scan] that is far less costly and labor intensive than unloading the cargo."[126]

MT may help detect SNM in other applications. The Coast Guard is concerned that small boats could enter a port carrying a nuclear bomb.[127] This is of particular concern for Miami, which small boats could reach from the Bahamas. Because muons penetrate about 15 meters of water, it might be possible to construct a muon detection system for boats, with an array of detectors above the water and several meters under water to detect entry and exit tracks of muons. Boats would stop in this array, perhaps for several minutes, while a tomographic image was built up. This application appears scientifically feasible, though engineering it would be considerably more difficult than for a land-based system. Another application might be detection of stowaways or contraband by locating suspicious voids in medium-Z material.

MT would be of use in detecting radiological material, such as might be used to make a "dirty bomb," or radiological dispersal device. Such materials are not high Z. Two such materials often mentioned are cobalt-60 and cesium-137. Cobalt has a Z of 27, and cesium has a Z of 55. However, because of their intense radioactivity, a large amount of dense shielding, such as lead, would be required to block gamma rays. For example, it would take a lead sphere two feet in diameter to shield most of the gammas from 0.11 cc of cobalt-60; MT, like existing x-ray systems, could readily detect such shielding.

Scanning Cargo or Analyzing a Terrorist Nuclear Weapon with Nuclear Resonance Fluorescence[128]

Two problems

Nuclear resonance fluorescence (NRF), described below, seeks to detect SNM in containers. At issue: Is NRF a useful approach for this task?

NRF may address a second problem. Discovery of a nuclear weapon in a cargo container would require an urgent effort to disable it and to gather forensic data. Both efforts would benefit from detailed information about the weapon's design. Several techniques can provide such information. Radiography or MT can show the shape and location of components; discovering that the weapon had a thermonuclear stage, for example, would show that it was manufactured by a nation and

---

[126] Information provided by Lawrence Delaney, Senior Vice President for System Development, Decision Sciences Corporation, e-mail, June 30, 2009.

[127] See "Feds Fight Threat of Small-Boat Terror Strikes," CNN, April 27, 2008, at http://www.cnn.com/2008/US/04/27/small.boat.terror.ap/index html#cnnSTCText.

[128] Stephen Korbly, Director of Science, Passport Systems, Dennis McNabb, Deputy Division Leader, N Division, Lawrence Livermore National Laboratory, and Glen Warren, Senior Research Scientist, Pacific Northwest National Laboratory, provided detailed information for this section, April-August 2008. Others have commented as well to provide alternative perspectives.

could have much higher explosive yield than a terrorist-made bomb. Interrogation using neutrons or high-energy gamma rays can provide information about SNM. NRF may be able to provide different types of data. Knowing what kind of chemical explosive the weapon contained, or combining information on location of electronics with information on their chemical composition, or knowing the mix of isotopes and impurities in SNM, would aid in dismantlement or might point to the source of the weapon. At issue: Is NRF a useful approach to determining the materials in a weapon?

Background

When atoms of a given element are illuminated with photons above an energy threshold unique to that element, their electrons absorb the photons' energy and move to a higher energy level, a so-called "excited" state. The electrons then drop back to their normal state, emitting photons that are slightly less energetic than the inbound photons. For example, certain elements or minerals illuminated with ultraviolet light (which is more energetic than visible light but less so than gamma rays) give off visible light.[129] This emission of light is called fluorescence.

A different type of fluorescence provides more detailed information. Each isotope has a unique combination of numbers of protons and neutrons in its nucleus, so it vibrates at unique frequencies (the resonant frequencies). When the nucleus is struck by a photon at precisely that energy level—sometimes to within a few hundredths of an electron volt in a beam with photon energies spread over a range of 1 MeV or more—it will absorb the photon and move to an excited state. The nucleus then reverts to its initial state, giving off photons very slightly less energetic than those that it absorbed. This process is known as nuclear resonance fluorescence, or NRF. NRF produces a gamma-ray spectrum unique to each isotope (though different than the gamma-ray spectrum produced by radioactive decay). Identifying the spectrum of the emitted photons identifies the element and isotope. This is of particular importance in differentiating between fissile U-235, which can be made into a nuclear weapon, and non-fissile U-238, which cannot. Unlike the use of neutrons or high-energy photons to stimulate the emission of neutrons or photons in SNM, NRF causes fluorescence in almost all isotopes of elements with Z>2 (helium), so it can identify a wide range of materials, not just SNM and other radioactive isotopes. For example, if nuclear material is shielded by lead, the identification of the various lead isotopes and their ratios may provide information as to where the lead was mined. Technical experts consulted for this report were aware of no other technology that permits identification of a weapon's materials without opening the weapon. CBP could also use NRF to identify other contraband and to check customs manifests.

Absorption of photons creates an additional signature. X-rays are generated in a broad spectrum of energies without sharp peaks. If a beam of such photons is sent through a cargo container or other object, and the detector on the other side can record the energy of each transmitted photon, a hole or "notch" in the spectrum at a certain energy level means that some particular material has absorbed photons at that energy level through NRF and then subsequently re-emitted the absorbed photons at about the same energy level. Since NRF photons are emitted in all directions, only a small fraction of them reach the detector. The energy level of the notch indicates what material is present. For example, the notch for U-235 occurs at 1.73 MeV.

---

[129] For an image of minerals fluorescing under ultraviolet light, see Glenbow Museum, Calgary, Alberta, "Fluorescent Minerals," at http://www.glenbow.org/collections/museum/minerals/flourescent.cfm.

Technology description

To detect NRF, an accelerator generates a beam of x-rays, a photon detector records the radiation spectrum generated by the material being interrogated, and an algorithm matches NRF peaks against a library of such peaks. Photons resulting from NRF are differentiated from the incoming photon beam because the latter produces a broad continuum of photon energies while photons generated by NRF produce very narrow peaks.[130] Further, the photon beam travels in a forward direction, while the NRF signal is emitted in all directions, so a photon detector placed behind and to the side of the material being interrogated (relative to the direction of the photon beam) detects photons traveling backward from the beam direction, which are mainly NRF photons. **Figure 18** illustrates this geometry.

Figure 18. Schematic Diagram of Nuclear Resonance Fluorescence System

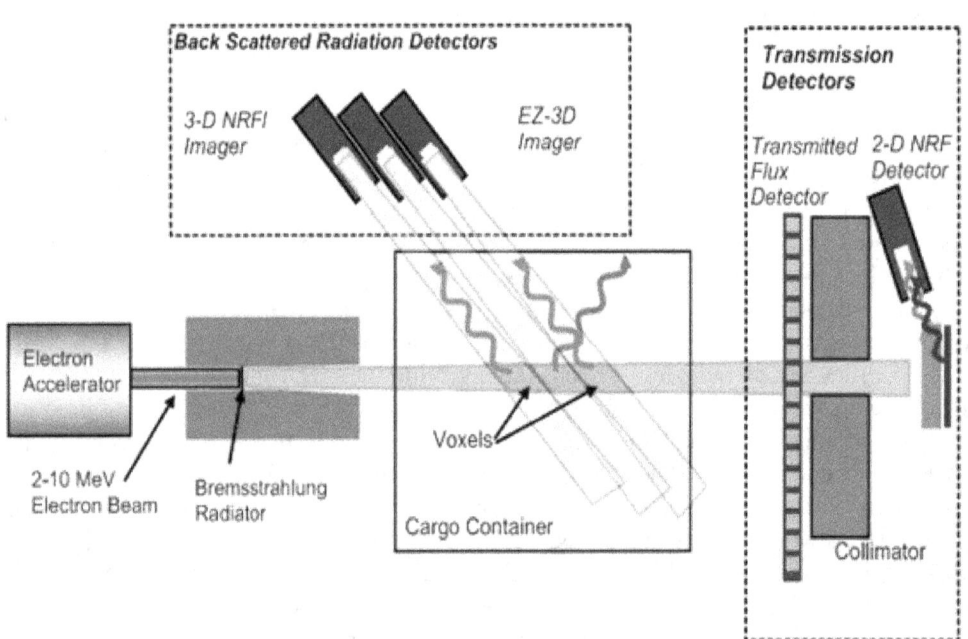

Source: Passport Systems, Inc.

Notes: From left to right: An accelerator generates a beam of electrons of 2 to 10 MeV. They strike a target (bremsstrahlung radiator), generating photons. A piece of high-Z metal allows only those photons moving toward the cargo container to exit. Photons pass through the container and induce nuclear resonance fluorescence, which releases photons of specific energies. NRF imagers "view" the container. The intersection of the photon beam and the "view" of each imager forms a voxel. Imagers are placed and aimed so that they view only photons moving backward from the direction of the beam in order to maximize the number of NRF-induced photons while minimizing the number of beam photons that reach the detectors. The system uses different types of detectors on the far side of the container.

DNDO is studying another approach to NRF using a beam of photons having a single energy level ("monoenergetic photons") near that needed to induce NRF in a particular isotope. At present, that method is technically difficult, costly, and requires large and delicate equipment,

---

[130] For a description of this process, see Passport Systems, Inc., "Technology," at http://www.passportsystems.com/tech htm.

making it unwieldy for deployment in the field. However, according to DNDO, this method has been verified experimentally and the Stanford Linear Accelerator Center has demonstrated one type of accelerator technology "that might lead to a mono-energy photon source that could be compacted into a 20-foot cargo container."[131]

Passport Systems, Inc., is developing an NRF imaging system, Passport MAX (Material Advanced Inspection), under contract to DNDO to detect SNM in cargo containers. It uses a commercial electron accelerator with a beam that can be varied from 2 to 10 MeV, depending on the materials and containers being searched, to produce a photon beam with energies ranging from several hundred keV to the maximum energy of the electron beam. The beam is collimated;[132] as it scans a container, it excites nuclei in its path that emit photons. A germanium gamma-ray detector views the emitted photons scattered backwards from one region at a time, records their energies, and constructs a spectrum. The intersection of the collimated beam with the detector's view creates a voxel, and the spectrum shows the type and quantity of each isotope in that voxel. An algorithm constructs a three-dimensional image of the container's contents. Passport MAX would include other detector subsystems as well: EZ-3D, as described under AS&E CAARS; a radiography imager; and an NRF detector that detects notches in the transmitted photon spectrum. Passport Systems indicates that this approach could also be used to scan smaller items such as an aircraft cargo container[133] or a terrorist nuclear weapon.[134]

Because the NRF imaging component can examine only one region of interest at a time, the CONOPS envisions using the other components of Passport MAX to locate volumes of interest, and then using NRF to interrogate them. According to Passport Systems, "the complete system would scan a 40 ft container for SNM in an average of about 15 seconds. If there were indeed SNM in a container it may take longer (minutes) to identify the material as SNM. However, we anticipate that the actual number of containers with SNM would be very small."[135]

Potential benefits

For detecting a terrorist nuclear weapon or SNM: (1) The system would identify each isotope, and would alarm on threat substances, with no operator input required. (2) While GADRAS must account for all the spectral data, the algorithm required for the Passport system need only account for spectral peaks, making for a simpler algorithm. (3) The system is to identify most isotopes that CBP finds in contraband, eliminating some false alarms that occur with radiation portal monitors, such as from radioactive potassium. (4) An NRF-based system can scan a cargo container quickly for SNM and high-Z shielding material. The average scan rate for the EZ-3D mode is 15 seconds, but the system would automatically adjust the speed at which individual

---

[131] Personal communication, Dr. R. Leon Feinstein, DNDO, August 8, 2008. This technology would accelerate particles to high energies over much shorter distances than are possible at present. For example, existing accelerators can increase the energy of particles (e.g., electrons) by 5 to 10 MeV per meter; the new technology might increase that to over 150 MeV per meter, making for a much more compact accelerator.

[132] Collimation filters many forms of electromagnetic radiation so that only photons traveling in a certain direction are allowed through. In the case of x-rays or gamma rays, a collimator is typically a plate of lead or tungsten with many small parallel holes drilled through it.

[133] William Bertozzi and Robert Ledoux, "Nuclear Resonance Fluorescence Imaging in Non-Intrusive Cargo Inspection," *Nuclear Instruments and Methods,* B241, 820 (2005), p. 7.

[134] Personal communication, Stephen Korbly, Passport Systems, June 9, 2008.

[135] Personal communication, Stephen Korbly, Passport Systems, August 8, 2008.

containers are scanned. If it detected little attenuation of the beam, it would scan faster. Conversely, if EZ-3D identified a region of interest, the system would reposition the container to analyze that region using NRF and photofission signatures and would likely increase scan time for that container. The impact of such delays on average throughput rate would depend on such factors as CONOPS and number of anomalies in containers being scanned.[136] (5) Passport Systems states that laboratory experiments and simulations predict that Passport MAX will be able to meet the scanning requirement that DNDO has set for CAARS, i.e., that it would have a 90% probability of detecting 100 cc of high-Z material, and a false alarm probability less than 3%, both with 95% confidence.[137]

For characterizing a nuclear weapon: (1) This system could determine the composition of a nuclear weapon. These data would be of value for disabling a weapon and for nuclear forensics. (2) An NRF-based system can identify the isotopic composition of uranium or plutonium and any impurities, which would be of value in nuclear forensics. (3) A potential difficulty with radiation detection is that large quantities of shielding in a container may block photons or neutrons as they enter and leave the container. The shielding problem would diminish if this system were used against an already-identified terrorist nuclear weapon because the weapon would be shielded only by its casing. (4) A *bremsstrahlung* source generates photons over a wide range of energies, and detectors are able to record a similarly wide range of emitted photons. As a result, the system could identify multiple materials quickly.

Status, schedule, and funding

From 2004 to 2008, DHS awarded Passport Systems several contracts totaling $8.4 million to build a proof-of-concept (PoC) scanner that is fully integrated and functional. In 2005, the contract was transferred from the DHS Homeland Security Advanced Research Projects Agency to DNDO for management. According to Feinstein,

> The NRF PoC system demonstration and evaluation completed on August 4-6, 2008. The primary purpose of this PoC test is to demonstrate full functionality and automation. This requires all critical components to operate as specified in an integrated architecture similar to an operational scanner. The Passport PoC subscale system successfully demonstrated its ability to automatically select high-Z [regions of interest] with EZ-3D and auto-identify the isotopic content of [these regions] with NRF. This was accomplished with a variety of cargo and with a mixed set of high-Z material and contraband. Other NRF applications are still being explored including cargo manifest-checking and forensics.[138]

Most of the major hardware components that the system uses—accelerator, detector, computer, and display—are commercially available. Passport Systems estimates that its system will be available for commercial delivery by mid-2010 at a unit cost of $5 million to $10 million depending on system configuration. Feinstein states, however, that that system "will not have completed the DHS phased-milestones of development, testing, evaluation and cost-benefit analysis" by that time. He further states that DNDO is developing enabling technologies, such as improved accelerators and detectors, "that, if successful, could significantly reduce the overall

---

[136] In order to identify the composition of a voxel with high confidence, the germanium detector must receive enough photons. If the material is dense, a longer scan time may be needed to accumulate the required number of photons.

[137] Personal communication, Stephen Korbly, Passport Systems, Inc., August 23, 2008.

[138] Personal communication, August 9, 2008.

size and cost [of the NRF system] (by more than a factor of two) as well as improve its performance and speed," though such technologies would not likely be ready for use in a commercial system by mid-2010.[139]

In September 2008, DNDO awarded Passport Systems a contract worth up to $9.3 million to build a full-scale prototype unit for an Advanced Technology Demonstration (ATD) of the NRF system under DNDO's Shielded Nuclear Alarm Resolution program.[140] This contract runs for 2½ years. The ATD scanner is intended to demonstrate performance on cargo containers and is the continuation of the previously demonstrated proof-of-concept (PoC) installation built by Passport at the University of California at Santa Barbara (UCSB), also under DNDO contract.

Many organizations have been conducting research into NRF to address national and homeland security issues since 2004. Lawrence Livermore National Laboratory (LLNL), Pacific Northwest National Laboratory (PNNL), and Passport Systems have collaborated to characterize the NRF response of U-235 and of Pu-239. PNNL has developed a computer model of NRF and has used simulations from it to reproduce results from laboratory measurements;[141] Los Alamos National Laboratory is developing another such model.[142] Duke University, Idaho National Laboratory, Idaho State University, LLNL, PNNL, Passport Systems, Purdue University, and University of California at Berkeley are conducting basic research on NRF. Passport Systems built its proof-of-concept prototype at University of California at Santa Barbara in order to use the accelerator in that university's free electron laser facility, and PNNL and Duke University conducted NRF measurements on U-238 and Pb-208 using the High Intensity Gamma Source at Duke as a source of nearly-single-energy photons.

Risks and concerns

Scientific risks and concerns

(1) A key scientific task is to conduct more experiments to identify the energy levels at which materials of interest undergo NRF. In particular, it would be useful to measure NRF spectra of isotopes of uranium and plutonium other than U-235 and Pu-239, and to measure spectra of materials used in nuclear weapons of other nations (e.g., for alloys) for purposes of nuclear forensics. (2) Another task is to develop the algorithms to identify materials quickly based on their NRF energies. (3) The Passport MAX geometry is designed to improve the signal-to-noise ratio by focusing on photons scattered backwards as a result of NRF, thus avoiding most of the photons generated by the interrogation beam, which travel in a forward direction. However, beam photons may have an energy range of nearly 10 MeV, while the energy range of photons that produce NRF in a particular isotope may be only a few hundredths of an electron volt. As a result, only one out of a few hundred million photons may have an energy level that produces NRF in

---

[139] Personal communications, August 9 and 21, 2008.

[140] See U.S. Department of Homeland Security. "Advanced Technology Demonstration for Shielded Nuclear Alarm Resolution," Broad Agency Announcement 08-102 for the Domestic Nuclear Detection Office, Transformational and Applied Research Directorate, March 2008, HSHQDC-08-R-00020, https://www.fbo.gov/download/6f6/ 6f6f82cc03fe6fcbf298a7e3903a15b7/HSHQDC-08-R-00020(3-28-08).doc.

[141] David Jordan and Glen Warren, "Simulation of Nuclear Resonance Fluorescence in Geant4," in Institute of Electrical and Electronics Engineers, Nuclear Science Symposium Conference Record, 2007, vol. 2, pp. 1185-1190.

[142] This model uses the Monte Carlo N-Particle Transport Code, developed by Los Alamos National Laboratory; see http://mcnp-green.lanl.gov/.

that isotope, making it hard to detect NRF-generated photons. On the other hand, if the NRF system is tuned correctly, very few photons of the same energy as the NRF photons scatter backwards to the detector, facilitating detection by increasing signal-to-noise ratio. (4) While NRF has routinely been used to detect gram samples, the limitations of the detectable mass of various isotopes has apparently not been quantified, so it is not clear that NRF can be used to detect microgram (or smaller) quantities of all isotopes, possibly limiting the applicability of NRF to nuclear forensics.

Engineering risks and concerns

(1) More work is needed to develop the detection algorithms. (2) The accelerator for this system, called a Rhodotron,[143] is built to order, and it may take the manufacturer a year to build one;[144] unless the manufacturer can build them much faster, it would be difficult to procure Passport MAX units in quantity. As of February 2010, Passport Systems was building a prototype of a high duty cycle accelerator[145] that could be produced more rapidly. In addition, since the accelerator footprint would be much smaller than that of the Rhodotron, Passport MAX would become smaller, which Passport Systems argues would make it or other NRF-based scanners much more attractive for mobile operations. (3) Once the tasks noted under "scientific risks and concerns" are completed, an engineering task would be to integrate hardware and software into an operational system, requiring many tradeoffs between cost, performance, and schedule. This may be particularly complicated for Passport MAX given the many separate detector units it uses, as shown in **Figure 18**. A demonstration of the full-scale ATD scanner under the Shielded Nuclear Alarm Resolution program is scheduled for 2010-2011; if successful, it would significantly reduce this risk.

Cost and schedule risks and concerns

The projected unit cost, $5 million to $10 million, is at a level that might preclude ordering large numbers of units. Regarding schedule, as of 2008 it was difficult to predict when an early-stage development program would become commercially available given the work that remained to be done and the possibility of unanticipated problems.

As of March 2010, Passport MAX had progressed from early-stage development to Advanced Technology Demonstration (ATD), with construction of the ATD unit scheduled to begin in 2010. As it has progressed, the amount of work remaining and the range of potential problems have decreased, so risk to schedule has decreased as well. Also as of March 2010, DNDO continued to fund enabling technologies, such as low-cost/high-resolution detectors and high duty cycle accelerators, that may, if successful, significantly reduce the cost and size of the commercial system and increase its NRF performance speed. As a result, Passport Systems stated that if the

---

[143] The Rhodotron is made by Ion Beam Applications, a Belgian company. The company states, "The Rhodotron® is a recirculating accelerator where electrons gain energy by crossing a coaxial-shaped accelerating cavity several times. This original design makes it possible to operate the machine in continuous mode for maximum efficiency and throughput." http://www.iba.be/industrial/rhodo-files/rhodo.php.

[144] Information provided by Ion Beam Applications, October 24, 2008.

[145] "Duty cycle" refers to the fraction of time that an accelerator's beam is on. Most accelerators operate at duty cycles of about 0.1 percent, i.e., the beam is on only 1/1000 of the time. Passport Systems states that its accelerators use a beam with duty cycles in the range of 5 percent to 100 percent. Information provided by Stephen Korbly, Passport Systems, e-mail, March 1, 2010.

MAX ATD scanner demonstration is successful in 2010, the company could deliver a commercial version of that system in 2011.[146]

Operational risks and concerns

The system would need to be large enough to hold a tractor-trailer truck. The Rhodotron used in the Passport MAX generates a considerable amount of radiation that requires containment. To meet a radiation safety requirement of no standoff zone, the system is completely enclosed, and a preliminary estimate by Passport is that it would be 90 to 100 ft long; 20 to 30 ft wide at its widest point, where the detection equipment is located; and several stories high at that point. These dimensions are dictated by the requirement to scan a container that is 40 ft long, 9 ft wide and 14.5 ft high. A system for smaller objects would be correspondingly smaller. Passport states that the enclosure will reduce the radiation dose outside the system to levels low enough to not require an exclusion zone. On the other hand, the system is quite large using existing commercial off-the-shelf technologies, which could be a problem in ports where space is at a premium.

Potential gains by increased funding

Passport Systems, a small company, does not have some key equipment of its own; for example, it is using an accelerator at the University of California at Santa Barbara as the photon source for its prototype. Added funds, Passport says, would enable it to buy needed equipment, advance supporting technologies, and hire more staff for engineering, algorithm development, and manufacturing. The prototype Passport MAX uses a considerable amount of expensive off-the-shelf hardware; with added funds, Passport says it could design less costly components specifically for use in its system.

Potential synergisms and related applications

(1) NRF could help identify illicit cargo in addition to SNM, such as explosives or chemical weapons. (2) NRF could help verify the manifest (list of contents) of a cargo container. (3) NRF might be used in nuclear forensics to identify rapidly the materials present in radioactive debris from the detonation of a terrorist nuclear weapon. (4) In nuclear nonproliferation applications, it could be used to analyze the isotopic composition of spent nuclear fuel. (5) New detector material could improve the sensitivity and resolution of the system, reducing the amount of electrical energy needed to run it and the amount of shielding needed, thereby reducing acquisition and operational costs.

---

[146] Information provided by Stephen Korbly, Passport Systems, e-mail, March 1, 2010.

Detecting SNM at a Distance[147]

The problem

It is hard enough to detect SNM at a range of several meters. Yet the ability to detect SNM or nuclear weapons at a kilometer or more, or in territory to which access is denied, would have great value in the fight against nuclear terrorism, such as locating SNM along a smuggling route, in a distant vehicle or facility, or in ships at sea, or finding a terrorist nuclear bomb in a city. This capability would also help protect military forces by enabling detection of terrorist attempts to use a nuclear weapon to destroy a military base or a carrier battle group. Making the task harder still, SNM might be shielded, and these missions would require a high search rate because of the need to scan a large area quickly or because SNM might be visible to the system for only a brief time. A system to perform this mission would be of value. At issue: Can a system be designed and deployed to detect SNM at an operationally useful standoff distance and search rate?

Background

The Defense Threat Reduction Agency (DTRA), a unit of DOD, is pursuing capabilities to increase the distance at which SNM can be detected, with the goal of increasing it to 1 km or more. DTRA states its rationale as follows:

> DTRA/DoD's motivation for pursuing Active Interrogation arises from the context in which a search for nuclear material would be conducted—potentially in a hostile environment over large areas (due to limited intelligence pinpointing the exact location of the material). DTRA/DoD is also challenged by the need to make the deployed equipment robust/well engineered enough to survive a range of harsh environments which is different from equipment use in a static mode at a border checkpoint.[148]

A scientific panel said, "Radiation attenuation due to shielding is an exponential process and so even moderate amounts of shielding can have significant effects. At 10 metres, the radiation emissions of shielded gamma ray and neutron sources are at, or below, natural background rates in almost all cases."[149] DTRA states, "The only way to overcome this physical reality is to stimulate the radiation emitted by SNM to a level many times above background. This can be done, for example, by using a beam of high energy photons to artificially induce photofission, and then detecting the resulting fission signatures. Beams of other types of radiation also have the potential to increase these detectable signatures through other reactions with the SNM."[150] As of April 2010, DTRA was continuing its investigation of the use of protons for active detection, including how a proton beam interacts with various materials and how to integrate passive detection and imaging methods with active interrogation beams.

---

[147] Dr. G. Peter Nanos, Jr., Associate Director for Research and Development, Defense Threat Reduction Agency (DTRA), Major Brad Beatty, USAF, Branch Chief, Standoff Detection Branch, Nuclear Detection Technology Division, DTRA, and Dr. Luc Murphy, Research Scientist, Locate and ID Branch, Nuclear Detection Technology Division, DTRA, and others at DTRA provided detailed information for this section, personal communications, April-July 2008. Others have commented as well to provide alternative perspectives.

[148] Personal communication, August 5, 2008.

[149] The Royal Society, "Detecting Nuclear and Radiological Materials," RS policy document 07/08, March 2008, p. 5, available at http://royalsociety.org/displaypagedoc.asp?id=29187.

[150] Personal communication, August 5, 2008.

Technology description

DTRA is sponsoring several remote-detection systems. This section considers the Photonuclear[151] Inspection and Threat Assessment System (PITAS), a project funded by DTRA and conducted by Idaho National Laboratory (INL). In 2008, PITAS was DTRA's remote-detection system closest to deployment; in April 2010 DTRA stated, "In FY10, a project was initiated to build the Integrated Standoff Inspection System (ISIS) to provide a robust system for the standoff detection of special nuclear material. This effort builds on the successful work of the PITAS project, while allowing the PITAS to continue to support experiments in the area of active detection." Data from PITAS will "allow a comparison of experimental results with simulation results."[152] In April 2010, DNDO awarded Raytheon a contract for $20.5 million for R&D on ISIS;[153] DNDO made no other awards for this effort.[154] Minimum requirements ("threshold") for ISIS include distance from accelerator to target greater than 100 m, with a goal of 1,000 m; distance from target to detector greater than 50 m, with a goal of 500 m; detection time less than 10 min, with a goal of less than 1 min; and maximum weight 8 tons and transportable by commercial aircraft, with a goal that one helicopter could transport the system.[155] In May 2010, DNDO stated that the current timeline calls for a demonstration of ISIS in FY2012, and that it has no estimate of the schedule beyond then.[156] Because the contract for ISIS R&D was awarded in April 2010, little information is available on that program. Further, as noted, ISIS builds on the work for PITAS. Accordingly, this section continues to focus on PITAS.

Figure 19 shows schematic and cutaway views. The main hardware components are an accelerator and detectors. PITAS would use a powerful linear accelerator, capable of generating a 30-MeV electron beam, to create x-rays that would be aimed at the target. These x-rays have a range of energies, with the maximum equal to that of the electron beam. Air attenuates lower-energy x-rays in the beam, so the target would be struck by x-rays well above the energy needed to induce photofission. Detectors might be located in the same unit as the accelerator. However, radiation spreads out (and diminishes in intensity) as the square of the distance, and the atmosphere would absorb many neutrons and gamma rays, so detection can be more effective if the detectors are separated from the accelerator and placed near the target. For example, detectors might be placed next to a smuggling route or on unmanned aerial vehicles with the accelerator some distance away. The x-ray beam would be used in a pulsed mode with the detectors attempting to detect SNM signatures, such as delayed neutrons and gamma rays (see **Appendix**), in the intervals when the beam is off.[157]

---

[151] "Photonuclear" refers to a nuclear reaction caused by a photon, in this case fission of SNM induced by high-energy photons.

[152] Information provided by DTRA, e-mail, April 22, 2010.

[153] "Raytheon Awarded Contract for Integrated Standoff Inspection System," Raytheon news release, April 26, 2010, http://raytheon.mediaroom.com/index.php?s=43&item=1546&pagetemplate=release.

[154] Information provided by DTRA, e-mail, April 30, 2010.

[155] U.S. Department of Defense. Defense Threat Reduction Agency. Broad Agency Announcement HDTRA1-09-NTD-BAA, "Advanced Detector Development (ADD) and Nuclear Forensics Research and Development Programs," November 2008, pp. 43-44, https://www.fbo.gov/download/4f8/4f898fe516d2145e703eaf8bbd33ff12/NTD-09-BAA.pdf.

[156] Information provided by DTRA, e-mail, May 10, 2010.

[157] Data cannot be gathered when the beam is on, for there would be far more photons from the beam than photons from fission of SNM, making it difficult or impossible for the detectors to identify the latter.

Figure 19. Schematic Diagram of Photonuclear Inspection and Threat Assessment System (PITAS)

Source: Defense Threat Reduction Agency.

Notes: The top image shows the concept of operations for PITAS. A "monocentric" detector is one that is located with the PITAS unit; a "bicentric" detector is one that is located elsewhere, typically near the object being inspected. The bottom image is a cutaway view. The accelerator is on the left; it generates bremsstrahlung photons. The refrigerator-shaped object is the power supply and associated data collection systems, and the object on the far right is the power converter and timing circuits. The detector unit is not shown but could be included in the container housing the PITAS unit.

Potential benefits

If PITAS or a follow-on system like ISIS works as anticipated, it would enable the United States to detect SNM at standoff range for the first time. This capability could be used for counterproliferation, counterterrorism, and force protection.

Status, schedule, and funding

As of April 2010, PITAS is to remain available as an experimental tool for at least the next few years. It will be used to address issues associated with field use of accelerator-based technologies, such as ruggedness and reliability. There are plans to improve PITAS diagnostic equipment in 2010 to support research. Funding levels for experiments with PITAS are not publicly releasable.

Risks and concerns

Scientific risks and concerns

(1) By one calculation, even a collimated x-ray beam could not place enough x-ray photons on a small piece of SNM in the target (such as a cargo container) to induce a detectable set of unique signals under ideal conditions.[158] Trying to detect SNM at a distance under real-world conditions, such as a container being towed by a truck along a winding road, would be even more difficult. (2) How effective would shielding be at stopping inbound x-rays or protons and outbound neutrons and gamma rays resulting from fission? (3) Gamma rays and neutrons resulting from fission of SNM would radiate in all directions, spreading out with distance. A source that caused 1,000 photons per second to strike a detector 1 meter square at a distance of 1 meter would cause only 1 photon to strike that same detector every 1,000 seconds at a distance of 1 km—disregarding attenuation by the air.[159] Yet enough photons must strike the detector to be detected and to be differentiated from background sources.[160] Regarding neutrons, fission generates many fewer neutrons than gamma rays, and the low-Z atoms in the air would attenuate neutrons strongly. Would enough gamma rays and neutrons reach the detectors to be detected? Another concern is that very little research has been done on remote detection. As a result, the developers of PITAS would have to conduct field tests to determine what signals the beam generates in SNM and how to detect them, and would have to develop algorithms to process the data. As of April

---

[158] Jonathan Katz states: "In principle, a sufficiently long and narrow collimator could produce an arbitrarily narrow x-ray beam. The price paid is that there is very little energy in the beam. For example, if the initial divergence is 30 degrees (plausible for an x-ray beam), then a beam collimated to 10 cm in diameter at 200 m (5 X 10^-4 radian, or 2.5 X 10^-7 steradian) will contain about 2.5 X 10^-7 of the source's energy and power. The rest is absorbed in the collimator or scattered into a diffuse flux; those photons are of no use for detection. The result would be a tiny amount of energy on the target. Any signal from fission of SNM generated by that energy would be emitted roughly equally in all directions, so a 100 cm^2 detector collocated with an x-ray source 200 m away would only pick up about 2 X 10^-8 of the signal. If that same detector were 5 m from the target but the accelerator were 200 m from the target, the detector would pick up only 3 X 10^-5 of the signal. Of course, background radiation isn't reduced at all, and would be several orders of magnitude stronger than the signal from fission. These estimates also ignored attenuation in the air. For a 10-MeV photon (a typical product of a 30-MeV accelerator), the beam is reduced by a factor of about 3 every 250 meters due to attenuation in the air, and the attenuation of 1-MeV fission gammas is about 5 times as great (a factor of 3 every 50 m). The attenuation of fission neutrons is about a factor of 3 every 150 m." Personal communications, September 29-October 8, 2008.

[159] The surface area of a sphere is $4\pi r^2$, where r is the radius. The surface area of a sphere with a 1-meter radius is 12.6 square meters. The surface area of a sphere with a 1,000-meter radius is 12.6 million square meters. Thus a detector one meter square would receive one-millionth as many photons at a distance of 1 km as compared to a distance of 1 m.

[160] DTRA states, "It is recognized that perhaps the single greatest challenge in successful implementation of this technology is the placement of high sensitivity (and specificity) detectors at ranges that may have to be significantly closer than the interrogation to be effective. These studies, to include modeling and experimentation are proceeding as part of the research program." Personal communication, August 5, 2008.

2010, DTRA continues to use PITAS to investigate technologies (including radiation sources and detectors) to extract threat signatures from natural and beam-generated background.[161]

Engineering risks and concerns[162]

One candidate neutron detector for PITAS is helium-3 drift tubes, which have a thin metal wall and a wire stretched lengthwise down the middle of the tube. Would these tubes be rugged enough to deploy as part of a mobile system that might operate in difficult terrain and extremes of weather and temperature? More generally, can the entire system be made compact enough to operate under such conditions, yet sensitive enough to operate effectively? Since 2008, as noted earlier, it has become apparent that demand for helium-3 far exceeds supply. Accordingly, DTRA is considering other materials to detect neutrons, such as boron-10 and lithium-6.

Cost and schedule risks and concerns

With PITAS having transitioned to an experimental support program, cost and schedule risks and concerns are much less salient than would be the case for a deployment-oriented program. DTRA does not have any specific information to provide on this topic.

Operational risks and concerns

The accelerator would generate beams of very high energy, and might not be heavily shielded in many mobile applications. It thus poses some radiation risk to the operator. Could radiation exposure be made low enough? Whether the risk is deemed acceptable would depend on exposure, which in turn depends on radiation output, frequency of use, and shielding, and on the urgency of the mission. Regarding the latter point, detecting SNM that intelligence indicated was being smuggled along a known route for sale to terrorists would justify more exposure risk than would monitoring that route on a routine basis. Further development of accelerators might reduce the shielding required. DTRA remains interested in reducing the size of particle accelerator technology for ease of deployment. DTRA further notes, "Several National Council on Radiation Protection & Measurements (NCRP) scientific committees in which DTRA participates or monitors are currently convened (SC 1-18 and SC 1-19) to examine the use of ionizing radiation for the standoff detection of SNM."[163]

Potential gains by increased funding

Beyond PITAS, DTRA is contemplating another remote-detection system for which, it asserts, added funds would shorten the schedule by several years. That project would use a proton beam

---

[161] Information provided by DTRA, e-mail, April 22, 2010.

[162] For example, the American Association for the Advancement of Science held a workshop on helium-3, focusing on the shortage, on April 6, 2010; see http://cstsp.aaas.org/agenda_meeting.html for presentations made at the workshop.

[163] Information provided by DTRA, e-mail, April 22, 2010. Scientific Committee (SC) 1-18 of the National Council on Radiation Protection and Measurements is Use of Ionizing Radiation Screening Systems for Detection of Radioactive Materials That Could Represent a Threat to Homeland Security; SC 1-19 is Health Protection Issues Associated with Use of Active Detection Technology Security Systems for Detection of Radioactive Threat Materials. See National Council on Radiation Protection and Measurements, "Current Program," http://www.ncrponline.org/Current_Prog/Current_Program.html.

---

with very high energies. DTRA regards it as very promising because, it believes, these protons may be able to defeat shielding at 1 km. As of August 2009, DTRA maintained that this project would be paced by investment. DTRA stated that since most of the technical risk is at the start of the project, more funds would permit exploration of many paths simultaneously so that potential pitfalls and opportunities could be identified at an early stage. However, in May 2010, DTRA said that it "is still investigating use of a high energy proton beam as an active interrogation source, but there are no near-term projects which warrant accelerated funding at this point."[164]

Potential synergisms and related applications

PITAS or a follow-on system could be used in support of the Proliferation Support Initiative, which seeks to stop shipments of nuclear weapons and other weapons of mass destruction worldwide. For example, a system of this type could help search a ship suspected of carrying nuclear weapons, SNM, or uranium ore. Such systems might benefit from improved scintillator material for detecting radiation. They require a material with substantial (but not extreme) resolution that is rugged enough to deploy under harsh conditions and inexpensive enough to deploy a large panel in order to capture more of the signal. In 2010, DNDO issued a broad agency announcement seeking in part to develop such materials.[165]

Chapter 3. Observations

Observations on Progress in Detection Technology

Equipment commercially available at the time of the 9/11 attacks was limited in its capability. PVT radiation detectors could detect radiation but could not identify isotopes, and shielding SNM might defeat detection. Radiographic equipment could reveal dense objects, but relied on operator skill to flag potential threats. It might be possible to hide a nuclear artillery shell in a cargo of dense objects, and it would be difficult to pick out a small piece of SNM. Resolving alarms required time-consuming methods, such as using hand-held radioisotope identification devices or unpacking a container.

Capabilities of existing systems can be improved incrementally, such as by using different detector material, computers, algorithms, or CONOPS (e.g., scan time).

Systems now under development have the potential to reduce false positives (speeding the flow of commerce) and false negatives (improving security). Fission that neutrons or x-rays induce in SNM generates unambiguous signals. Dual-energy radiography detects high-Z material automatically. EZ-3D reveals high-Z material hidden in medium-Z material, and might be able to differentiate SNM from other high-Z material. These approaches detect useful signatures, but have drawbacks as well, such as low signal strength, complexity, high cost, or large size. The task is to utilize these signatures and minimize drawbacks in a system that can be fielded. Other

---

[164] Information provided by DTRA, e-mail, May 10, 2010.

[165] U.S. Department of Homeland Security. Domestic Nuclear Detection Office. "Advanced Radiation Monitoring Devices (ARMD): Near Term Research Project," Broad Agency Announcement (BAA) BAA10-DNDO-01, March 1, 2010, https://www.fbo.gov/download/ccf/ccf3ddfc085144aa0e5cc2cfbfa2cb65/ARMD_BAA_Final_2010.doc.

technologies, such as improved detector material and improved algorithms, also have the potential to improve detection capability.

It is difficult to predict the schedule of new detection technologies. In March 2008, the Royal Society, drawing on a workshop of experts, issued a report on nuclear detection that found, "In the medium term (5-10 years), there are promising opportunities to develop new technologies, such as muon detection systems. In the long term (10-20 years) detection could benefit from advances in nanotechnology and organic semiconductors."[166] In 2008, the company developing the muon tomography system thought the system could be commercially available in 2011. As of early 2010, that date had slipped to 2012 and the company had not passed its Test Readiness Review, a step to indicate whether a system is ready for its proof-of-concept demonstration. In 2008, some thought that nanocomposite scintillator technology could be available for transfer to industry by September 2009, but the project was canceled in January 2010.

It is difficult to evaluate prospects for R&D projects. Based on tracking the technologies presented in this report, it appears extremely difficult to evaluate how likely an R&D project is to succeed, even for the agencies that fund them, and one should not confuse a technical explanation and briefing slides with prospects for success. To succeed, a project must overcome many hurdles between concept and deployment. (1) The concept has to be scientifically sound. This is not always a given for projects that push the state of the art. (2) Even if scientifically sound, the underlying science must be transformed into a prototype through engineering. But materials may prove impossible to develop; laboratory-scale proof-of-concept equipment, where size and complexity are not a concern, may prove difficult to shrink in size; and algorithms may be unstable or may be confused by background radiation. (3) The prototype must be made into a system that is rugged enough to survive the bumps, vibrations, heat, cold, rain, humidity, dust, salt air, gasoline fumes, and whatever else people and nature may throw at it. (4) There must be a workable concept of operations: if it takes 1000 seconds to perform a scan, or if the false positive and false negative rates are too high, or if the operator cannot use the equipment easily, the equipment is useless. (5) The system must be affordable, however defined. It is hard to predict if a concept will make it past the next hurdle, let alone all five (and any others).

Here are several examples drawn from this report. Nanocomposite scintillators held the promise of being a gamma-ray detection material that would be sensitive, yet inexpensive and easy to produce on a large scale. Early research started in 2004, but DNDO and DTRA terminated the project in 2010. The AS&E CAARS project appeared promising, but encountered unspecified technical problems and DNDO terminated it; however, some of its technology is being applied to another project. Conversely, SAIC's CAARS depended on the development of an "interleaved" accelerator, one that could switch x-ray beams between two energy levels many times a second. An earlier attempt to develop such an accelerator failed, but SAIC's subcontractor, Accuray, was able to develop one that exceeded requirements by a substantial margin, contributing to the system's ability to differentiate among up to 15 bands of Z rather than simply indicating whether material in a cargo container was high-Z or not. This enhanced capability could help CBP agents search for contraband as well as SNM.

It is easier, less costly, and potentially more effective to accelerate a program in R&D than in production. DTRA believes that a significant increase in funding for proton beam technology,

---

[166] The Royal Society, "Detecting Nuclear and Radiological Materials," RS policy document 07/08, March 2008, p. 1, available at http://royalsociety.org/displaypagedoc.asp?id=29187.

a standoff detection technology in early R&D, might shorten time to deployment by several years by enabling researchers to consider many technical alternatives simultaneously to determine the most promising approach faster. It is hard to attain large schedule gains by accelerating production; such gains may entail high cost, such as multiple shifts or more production lines; and a rush to production may cause a project to fail. While R&D projects may also fail, more risk is tolerable in R&D because the investment is much less.

A modest amount of money spent in R&D can avoid looming problems. For example, GADRAS, a widely used algorithm for detecting SNM and other materials, runs on the standard Microsoft Windows operating system (OS) for personal computers. Microsoft introduces new generations of OSs from time to time. Typically, new OSs will support programs written for several generations of previous such systems. However, the Graphic User Interface (GUI) for GADRAS is written with the Visual Basic 6.0 compiler, which Microsoft no longer supports. At some point, Microsoft will likely introduce a new OS that will no longer support applications that are written with this compiler; GADRAS would then be unavailable to its users until it is updated. According to Dean Mitchell, who created GADRAS, updating that algorithm to run on current-generation OSs would avoid that problem, at a cost of perhaps $1 million a year for two years.[167]

R&D that leads to products that many systems can use may have a large impact on detection capability. Many detector systems have common elements—an accelerator, gamma-ray detector material, computers, algorithms—so improving any of these "building blocks" might improve the capability of many detector systems, including those in the field. Improved gamma-ray detector material can improve sensitivity, reduce cost, or both. An improved algorithm can boost performance. A more powerful computer permits the use of a more powerful algorithm, which may reduce false positives and false negatives.

On the other hand, it may not be possible to upgrade systems simply by swapping new components for old. Edward McKigney listed possible difficulties in the (hypothetical) case of upgrading systems by substituting higher- for lower-performance detector material:

> (1) Detector modules that cannot detect light with high efficiency would need to be redesigned. This is particularly relevant for existing portal monitors that use plastic scintillator material, where the optical design is poor. (2) Electronics for converting signals from detectors into data for algorithms ("readout electronics") that are not suitable for high-resolution readout and analysis, or are mismatched for the technology (such as if the old electronics read electrical charge while the new ones read optical signals), would have to be replaced. (3) Data analysis algorithms that cannot process signals from the new detector module would have to be replaced. (4) The volume of data from the new detector module might be greater than the existing algorithms, data transmission system or computers could handle, requiring new computers, algorithms, data transmission system, or some combination. (5) Electrical power systems would have to be changed if the power requirements for the old and upgraded systems did not match.

> So, at the extremes, it might be possible to upgrade only the detector module, or the only features of the old system that would remain after an upgrade would be the wide spot in the road and the guard shack. I would recommend that **the next generation of detector systems**

---

[167] Personal communication, July 19, 2008.

should be more modular so upgrades could be done while retaining as much of the value of deployed systems as possible.[168]

Synergisms may arise between technologies. Beams of neutrons or high-energy gamma rays used to induce fission in SNM may harm some products, expose stowaways to high doses of radiation, and require shielding to protect workers. Improved detector material and algorithms could lower the amount of radiation required for this technique, perhaps making it more usable for scanning containers.

Technical advances can place two systems in competition unexpectedly. Work is underway on several systems designed to induce fission as a way of detecting SNM. CAARS was not begun as a system of this sort. However, DNDO is investigating a technology add-on to give it that capability. If work proceeds on that path, CAARS and other such systems could be in competition.

Competition at the R&D level may be desirable. William Hagan, Assistant Director, Transformational and Applied Research, DNDO, states,

> if we can squeeze additional functionality out of a system, we want to do that. This will cause various approaches to be in competition for achieving a capability at the R&D stage but that is what we want to do so we can drive towards the most effective.
>
> More generally, I think that having multiple organizations pursuing the same R&D goal is a good thing because it allows for different approaches or more capable organizations to compete for the objective. This is a very effective mechanism in R&D. A classic recent example is the race to decode the human genome. Another is the race for commercial space flight. This kind of competition goes on all the time in the basic research community and I think we should encourage it. There is, of course, some limit to this, but we are far from that limit right now for radiation detection.[169]

The competitive position of systems in R&D may change over time. Technology development is dynamic. This report presents several examples. The SAIC CAARS overcame a key technical hurdle, the development of an interleaved accelerator, resulting in better performance than expected. The AS&E CAARS encountered problems that led to its termination. The Rapiscan Eagle, with an added algorithm to detect high-Z material, became a competitor to CAARS through the JINII program. Decision Sciences Corporation addressed problems with the original concept for its muon tomography scanner, such as using boron-10 instead of helium-3 in drift tubes because of the latter material's scarcity and designing a top/bottom scanner rather than a top/bottom/both sides scanner to make the footprint more compatible with traffic lanes at ports.

"Concept of operations" (CONOPS) is crucial to the effectiveness of detection systems. CONOPS details how a detection system would be operated to gather data and how the data would be used. Without it, a detection system would be valueless. Since CONOPS and systems are mutually dependent, the design of each must take the capabilities and limitations of the other into account.

---

[168] Personal communication, August 4 and 5, 2008; emphasis added.

[169] Personal communication, August 1 and 8, 2008.

Current equipment to detect and identify SNM makes use of two main signatures of this material, opacity for radiography to detect SNM, and gamma-ray emissions for spectroscopy to detect and identify SNM. However, as discussed in Chapter 2 and the **Appendix, SNM has many signatures in addition to opacity and gamma-ray emissions, and some systems under development attempt to make use of these other signatures.** If systems utilizing these other signatures were to be deployed, methods that might be used in an attempt to hide or mask opacity and gamma-ray signatures would not necessarily defeat these systems under development—complicating any terrorist attempts to smuggle nuclear weapons or SNM into the United States. At the same time, these future systems tend to be more costly and complex than current systems; whether the added benefits are worth the added costs is a political decision.

Detection systems have their limits. Systems to detect SNM at close range, such as at ports and land border crossings, are generally not applicable to detection of terrorists smuggling a weapon across a remote stretch of border. But that is not a flaw of the detection system. Detectors can work at "points," i.e., places where people or cargo may enter the United States legally. There, detectors attempt to find SNM or weapons that may be hidden in cargo. In contrast, at "lines," the vast distances between "points" along coasts or borders, any entry is illegal, so interdiction is a matter of law enforcement. Effective intrusion detection systems (TV cameras, seismic monitors) coupled with a CONOPS that provides rapid response may suffice, though they have a long way to go to become effective. At the same time, standoff radiation detection systems that have yet to be developed, mounted along borders at natural choke points or smuggling routes, might be of value for this mission.

Observations on Technical Progress and Congress

Congress has supported a broad portfolio of detection R&D projects that has created a pipeline with technologies expected to become available for operational systems from near-term to long-term. These systems exploit many signatures in addition to those of currently deployed systems, offering Congress the prospect of improved detection capability and a broader menu of choices. Several technical factors may influence the choice among technologies to support. For example: (1) Projects will advance at different rates. (2) Projects may benefit differently from an advance in a related technology. (3) As a project moves from research to development to deployment, cost and capability may vary from early projections.

Congress may wish to reevaluate current deployment decisions if it concludes that significantly more capable systems will be available in a year or two. Of course, any such decision would depend on comparing such factors as cost, footprint, ease of use, production rate, and the like for competing systems, and caution is necessary in assessing contending claims.

On the other hand, **it is difficult for Congress to choose among contending technologies.** Each requires evaluation in such terms as cost, scan time, ease of use, reliability, schedule, footprint, radiation exposure, spatial resolution, and ability to thwart shielding. Yet these data are difficult to obtain. Some are proprietary. Some are unknown: schedules may slip and costs rise, or technical advances may cause the opposite to occur. Developers of a technology tend to be its advocates, and see the strengths of their technology and a path to overcome its weaknesses. Even if these data can be obtained, it is necessary to weight data elements to support a choice among contending technologies. With many variables to be traded off against each other, how are weights to be assigned, and who decides? And can this weighting system function despite weaknesses in the data?

Congress has focused much attention on preventing terrorists from smuggling nuclear weapons or SNM into the United States in cargo containers. For example, P.L. 110-53, Implementing Recommendations of the 9/11 Commission Act of 2007, Section 1701, states, "A container that was loaded on a vessel in a foreign port shall not enter the United States (either directly or via a foreign port) unless the container was scanned by nonintrusive imaging equipment and radiation detection equipment at a foreign port before it was loaded on a vessel." While terrorists might attempt to smuggle in a nuclear weapon by other means, **developing technology to scan containers at seaports is a reasonable place to start.** Container-scanning technology can be modified for use in other situations, such as monitoring air cargo containers or passenger cars, which are easier to scan because they can contain much less shielding. Developing and deploying detection equipment for use at seaports ensures ruggedness and ease of use adequate for real-world applications, and forces governments at all levels to plan CONOPS.

More generally, some could argue that it is impossible to prevent terrorists from smuggling nuclear weapons into the United States, so there is no point in spending large sums in a futile effort. **Congress has rejected that approach, and has appropriated, in total, billions of dollars to deploy available systems and to support R&D on advanced technologies. Supporters of the R&D and deployment approach assert that it offers several advantages.**

- It has provided some capability quickly, increasing the odds of detecting weapons or SNM. An important example of this is the rapid deployment of passive radiation detectors to scan maritime cargo containers.

- This limited detection capability would help deter terrorists and would complicate plans to smuggle in weapons or SNM.

- Initial deployments provide data of use to subsequent deployments. They help refine what throughput, robustness, etc., front-line inspectors require of a system. They help refine CONOPS. They help define desirable features of an architecture. These results can make future technologies, systems, and architectures more effective.

- It has created an R&D pipeline that is intended to generate a steady stream of new technologies and systems.

- The resulting improvements in individual technologies, operations, and architectures can improve overall system effectiveness.

- As technologies become more capable, they can plug gaps in the current architecture. For example, remote detection might offer a way to monitor choke points in the United States or overseas that terrorists might pass through in transporting SNM or weapons.

Congress may wish to address gaps and synergisms in this portfolio. For example:

Gaps: Several systems may use helium-3 tubes for neutron detection, yet the supply is limited. Alternatives are available, but the longer developers take to switch to these alternatives, the longer it would take to deploy their systems because of the need to incorporate different detectors, modify algorithms, and test the revised system. Other gaps include sensors that can detect SNM at long range (e.g., over 100 m), sensors that can operate autonomously in remote areas, and large but inexpensive detectors that can distinguish SNM from other radioactive material.

Synergisms: A component, algorithm, or material developed for one system may be applicable to another. Projects are under way to develop more sensitive materials to detect gamma rays and neutrons. These materials can be used in systems that induce fission in SNM. Their improved sensitivity permits a smaller source (e.g., an accelerator) to generate the interrogation beam, reducing cost, complexity, and radiation dose. Similarly, if detector material offers only fair resolution of gamma ray spectra, then peaks in a spectrum may blur, requiring a complicated algorithm to deal with the uncertainties. Sharper resolution from improved materials would reduce these uncertainties, permitting simpler algorithms to be used. More powerful computers could support faster, more powerful algorithms, reducing scan time, false positives, and false negatives.

Minimizing gaps and maximizing synergisms have the potential to lead to more capable systems faster and at lower cost. Companies that considered using helium-3 for neutron detection might expedite deployment and reduce costs by sharing effort to develop an alternative neutron detector for their common use. Information on progress in developing more sensitive detector material would permit companies to incorporate such materials into their systems sooner, also speeding deployment and lowering costs. Is there a way that development could be shared or licensed so that companies, especially those working on government-funded projects, could avoid duplicating effort? And could this be done while retaining the benefits of competition?

In considering the Advanced Spectroscopic Portal, Congress and the Government Accountability Office examined in detail whether DNDO had followed proper procedures for testing competing systems. **An alternative means by which Congress could address testing is to direct the executive agency in charge of a system to conduct specified tests.** These tests would need to be designed, and perhaps overseen, by experts not affiliated with the relevant agency, company, or laboratory. Congress has ample access to the technical expertise required. The relevant congressional committees could consult with individual experts or with groups that have a long history of providing independent technical advice to the government, such as the American Association for the Advancement of Science, the JASON defense advisory group, the National Academy of Sciences, the National Council on Radiation Protection and Measurements, and the National Institute of Standards and Technology. In this way, Congress could seek a fair comparison between systems on variables of interest, such as scan times or the ability to detect specified targets in containers with specified cargoes, enhancing confidence in the test results and decisions based on them. Other alternatives exist. Congress could require DHS to establish an independent test and evaluation unit; obtain an outside review of DHS test and evaluation procedures; require DNDO to provide detailed reporting of each step in the acquisition process as it occurs; or provide for an external review of a program.

Observations on Technical Progress and Terrorism

Ongoing improvement in U.S. detection capabilities produces uncertainties for terrorists that seem likely to increase over time, adding another layer of deterrence beyond that of the capabilities themselves. Capability of fielded equipment may be upgraded. Terrorists may not know the capability or availability of future detectors. More advanced technologies should improve detection capability. It should be harder for terrorists to evade new systems than current ones. Detection may affect terrorists in another way. A nuclear weapon would be of immense value to them. Therefore, increasing the risk of detection would have a much greater deterrent effect for them than for drug smugglers, where detection and confiscation of drugs are part of the cost of doing business. The multiplication of technical obstacles to a successful terrorist attack may thus help deter an attack or an attempt to undertake a project to launch one.

At the same time, it is important to **avoid the "fallacy of the last move."** Herbert York, a former Director of Defense Research and Engineering, coined this term to argue that in the Cold War nuclear arms race, one side's actions typically led to countervailing actions by the other side.[170] The same principle applies to nuclear detection. This report suggests that some U.S. detection systems nearing readiness for deployment are more capable than current detectors. Yet if terrorists were to attempt to bring a nuclear weapon or SNM into the United States, they could use various techniques to evade detection by such systems, though these techniques might complicate the plot and increase the risk of detection by non-technical means. Further, the threat might increase in various ways, such as if new terrorist groups emerged or if more nations built nuclear power plants or nuclear weapons. For such reasons, Congress has funded, and executive agencies are pursuing, R&D with short- and longer-term time horizons. Also for such reasons, the global nuclear detection architecture may need to be updated from time to time. Thus, while the United States has an immense technical advantage in a competition of detection vs. evasion, and a pipeline of increasingly more capable technologies, it is important to recognize not only the dynamic aspects of advances in detection capabilities but also the dynamic aspects of the competition.

---

[170] Herbert York, *Race to Oblivion: A Participant's View of the Arms Race,* New York, Simon and Schuster, 1970, p. 211.

Appendix. The Physics of Nuclear Detection[171]

What Is to Be Detected?

Detectors must detect complete weapons, which can be quite small. During the Cold War, the United States made 155 mm and 8 inch (diameter) nuclear artillery shells. The United States made even smaller atomic demolition munitions, and there have been reports of Soviet-era "suitcase bombs." A weapon that terrorists fabricated without state assistance would surely be less sophisticated and, as a result, probably much larger. Detectors must also detect the types of uranium and plutonium used in nuclear weapons. The type of uranium used in weapons is harder to detect than plutonium because it emits much less radiation; it is also much easier to fabricate into a weapon component. It is important to detect small quantities of these materials in order to interdict stolen and smuggled materials because small quantities suffice to fuel a bomb. According to a widely quoted report by five nuclear weapon scientists from Los Alamos National Laboratory, it would take 26 kg of uranium, or 5 kg of plutonium (both of types discussed later) to fuel an atomic bomb.[172] These masses would fit into cubes 11.2 cm or 6.3 cm, respectively, on a side. The ability to detect even smaller masses would help thwart nuclear smuggling. How is it possible to find weapons or materials among the vast amount of cargo that reaches the United States each day? Fortunately, there are many clues.

Background

Photons

Nuclear detection makes extensive use of photons. Photons are packets of energy with no rest mass and no electrical charge. Electromagnetic radiation consists of photons, and may be measured as wavelength, frequency, or energy; for consistency, this report uses only energy, expressed in units of electron volts (eV).[173] Levels of energy commonly used in nuclear detection are thousands or millions of electron volts, keV and MeV, respectively. The electromagnetic spectrum ranges from radio waves (some of which have photon energies of 10^{-9} eV), through visible light (a few eV), to higher-energy x-rays (10 keV and up) and gamma rays (mostly 100 keV and up). An x-ray photon and a gamma-ray photon of the same energy are identical.

Gamma rays originate in processes in an atom's nucleus. Each chemical element has two or more isotopes. Isotopes of an element have the same number of electrons, and thus in most cases similar chemical properties, but different numbers of neutrons in their nuclei, and thus different nuclear properties. Each radioactive isotope emits gamma rays in a unique spectrum, a plot of

---

[171] John Valentine, Lawrence Livermore National Laboratory, provided invaluable assistance in explaining the science presented in this section, April-July 2008. Others reviewed and commented on this section as well.

[172] Mark et al., "Can Terrorists Build Nuclear Weapons?"

[173] An electron volt is a unit of energy used for measuring atomic and nuclear processes. One electron volt (eV) is equal to the amount of energy gained by a single unbound electron (one not part of an atom) when it accelerates through an electrostatic potential difference of one volt. It is equal to 1.6×10^{-19} Joules. For comparison, the energy release in the fission of one uranium atom is 200 million electron-volts, and the energy required to remove an electron from a hydrogen atom is 13.6 eV. Information provided by Defense Threat Reduction Agency, personal communication, August 5, 2008.

---

energy levels (horizontal axis) and number of gamma rays detected at each energy level (vertical axis). These spectra are a series of spikes at particular energy levels.[174] **Figure 1** and **Figure 2** show the spectra of uranium-235 and plutonium-239, respectively. Such spectra are the only way to identify an isotope outside a well-equipped laboratory. A detector with a form of "identify" or "spectrum" in its name, such as Advanced Spectroscopic Portal or radioactive isotope identification device, identifies isotopes by their gamma-ray spectra.

X-rays originate in interactions with an atom's electrons. Many detection systems use x-ray beams, which can have higher energies than gamma rays and thus are more penetrating. X-ray beams are often generated through the *bremsstrahlung* process, German for "braking radiation," which works as follows. An accelerator creates a magnetic field that accelerates charged particles, such as electrons, which slam into a target of heavy metal. When they slow or change direction as a result of interactions with atoms, they release energy as x-rays whose energy levels are distributed from near zero to the energy of the electron beam. They do not exhibit spectral peaks like gamma rays. This difference is important for detection.

Radioactivity

Radioactive atoms are unstable. They decay by emitting radiation, principally alpha particles (a helium nucleus consisting of two neutrons and two protons, thus having a double positive charge), beta particles (electrons or positrons, the latter being electrons with a positive charge), and gamma rays (high-energy photons). These forms of radiation are of differing relevance for detection. Alpha particles, being massive (on a subatomic scale) and electrically charged, are easily stopped, such as by a sheet of paper or an inch or two of air. Beta particles, while much lighter and faster, are also electrically charged and are stopped by a thin layer of material.[175] Gamma rays have no charge and can penetrate much more material than can alpha or beta particles. Depending on their energy, they may travel through several hundred feet of air. When an atom decays by emitting an alpha particle or beta particle, it transforms itself into a different element; it does not do so when it emits a gamma ray. Gamma-ray emission typically follows alpha or beta decay. As discussed in more detail below, each radioactive isotope that emits gamma rays does so in a spectrum of energies unique to that isotope. For example, the spectrum of U-235 has a prominent peak at 186 keV.

In addition to these typical means of radioactive decay, atoms of some heavy elements fission, or split into two smaller atoms. Of the naturally occurring isotopes, only U-238 spontaneously fissions with an appreciable rate (about 7 fissions per second per kg). One by-product of fission is the emission of neutrons (typically 2-3 neutrons per fission). Neutrons have no electrical charge and can penetrate dense materials, as well as many tens of meters of air.

---

[174] For the gamma ray spectra of various isotopes, see "Gamma-Ray Spectra of Isotopes," within the Radiochemistry Society website at http://www.radiochemistry.org/periodictable/gamma_spectra/. For the percentage distribution of the dozens of gamma rays from uranium-235, see "TORI Data" in "WWW Table of Radioactive Isotopes," available at http://ie.lbl.gov/toi/nuclide.asp?iZA=920235.

[175]

A half-centimeter of air will stop a beta particle emitted by tritium, while 0.04 cm of water or 2 mm of aluminum will stop a beta particle emitted by iodine-131. Eckhardt, "Ionizing Radiation—It's Everywhere," pp. 18, 19.

Fissile material

Some isotopes of heavy elements fission spontaneously or when struck by neutrons or high-energy photons, emitting neutrons and gamma rays in the process. U-235 and Pu-239 are unique in that neutrons of any energy can cause them to fission; they are called "fissile." Neutrons of much higher energies are required to cause other isotopes to fission. This characteristic of U-235 and Pu-239 allows them to support a nuclear chain reaction. Fissile material is essential for nuclear weapons; U-235 and Pu-239 are the standard fissile materials used in modern nuclear weapons. The Atomic Energy Act of 1954 designates them as "special nuclear material" (SNM).[176]

Plutonium is not found in nature. It is produced from uranium fuel rods in a nuclear reactor and is separated from uranium and other elements using chemical processes. Weapons-grade plutonium (WGPu) is at least 93% Pu-239. In contrast, uranium in nature consists of 99.3% U-238 and 0.7% U-235, with very small amounts of other isotopes. Enriching it in the isotope 235 for use as nuclear reactor fuel or in nuclear weapons cannot be done through chemical means because isotopes of an element are nearly chemically identical,[177] so other means must be used. For example, uranium may be converted to the gas uranium hexafluoride and placed in centrifuges specially designed to separate U-235 from U-238 based on the very slight differences in the weight of individual molecules. Uranium enriched to 20% in the isotope 235 is termed highly enriched uranium, or HEU; for use in nuclear weapons, uranium is typically enriched to 90% or so, though lower enrichments could be used. For purposes of this report, "HEU" is used to refer to uranium of 90% enrichment. HEU may also be produced from material that has been in a nuclear reactor. HEU produced in this manner contains small amounts of another isotope, U-232, which, as we shall see, is easier to detect than is U-235.

Detection

Nuclear detection uses neutrons and high-energy photons in various ways. Because they can penetrate different materials, they are the main forms of radiation by which most radioactive material can be detected passively, by "listening" for signals coming out of a container without sending signals in. Because of their penetrating properties, they can be used in an active mode to probe a container for dense material. X-rays or gamma rays are used for radiography, that is, creating an opacity map like a medical x-ray. Neutrons of any energy level, and photons above 6 million electron volts (MeV), can be shot into a container to induce fission in SNM. Fission results in the emission of neutrons and gamma rays, which can be detected. Gamma rays can also be used to identify a radioactive source. Neutrons, in contrast, do not have a characteristic energy spectrum by which an isotope can be identified, and it is difficult to measure their energy, though the presence of neutrons in certain situations, as discussed below, can indicate that SNM is present.

---

[176] Under the Atomic Energy Act of 1954, P.L. 83-703, 42 U.S.C. 2014, SNM is uranium enriched in the isotopes 233 or 235 or plutonium. The Nuclear Regulatory Commission has not declared any other material to be SNM even though the Act permits it to do so. U.S. Nuclear Regulatory Commission. "Special Nuclear Material." Available at http://www.nrc.gov/materials/sp-nucmaterials.html.

[177] The degree to which chemical properties of isotopes are similar "depends on the element. For hydrogen/deuterium the chemical differences are substantial; you cannot survive on heavy water. For other light elements they are small but produce measurable effects (the whole field of paleoclimatology is based on this). For uranium they are infinitesimal." Personal communication, Jonathan Katz, August 7, 2008.

Another characteristic of radioactive materials important for detection is the rate at which a material decays. The half-life of an isotope, or the time it takes for half the atoms in a sample to decay, is an indicator of the rate of decay, with shorter half-lives indicating faster decay. The half-lives of cobalt-60, plutonium-239, and uranium-235 are 5.3 years, 24,000 years, and 700 million years, respectively.[178] [179] Even if a source emits high-energy gamma rays, it will be difficult to detect if it emits only a few of them. Thus type, energy level, and quantity of radiation are important for detection.

Shielding and background radiation

Different materials attenuate neutrons and gamma rays in different ways. Heavy, dense materials like lead, tungsten, uranium, and plutonium have a high atomic number (the number of protons in the nucleus), or "Z." High-Z materials attenuate gamma rays efficiently.[180] In contrast, neutrons are stopped most efficiently by collisions with the nuclei of light atoms, with hydrogen being the most effective because it has about the same weight as neutrons.[181] The element with the nucleus closest in weight to a neutron is hydrogen, which in its most common isotope consists of one proton and one electron. Thus hydrogen-containing material like water, wood, plastic, or food are particularly efficient at stopping neutrons; other low-Z material is less efficient at stopping neutrons, but nonetheless more effective than high-Z material. Conversely, gamma rays are less attenuated by low-Z material and neutrons are less attenuated by high-Z material.

Different amounts of material are needed to attenuate gamma rays depending on their energy level. Gamma rays from WGPu are sufficiently energetic and plentiful that it is more difficult to shield WGPu than HEU. In contrast, as explained in the footnote, an inch of lead would render gamma rays from U-235 essentially undetectable, though as discussed later other uranium isotopes that may be present in HEU are more readily detectable.[182] Indeed, 186-keV gamma rays

---

[178] U.S. Department of Energy. Office of Environmental Management. Integrated Data Base Report—1996: U.S. Spent Nuclear Fuel and Radioactive Waste Inventories, Projections, and Characteristics, revision 13, December 1997; table B.1, "Characteristics of important radionuclides," http://web.em.doe.gov/idb97/tabb1.html.

[179] A more precise indicator of decay is specific activity, the number of curies per gram of material, where 1 curie = 3.7 x 10^10 disintegrations per second. Plutonium-241, for example, has a specific activity of 102 curies/gram, and its rapid radioactive decay makes it so hot that pieces of it glow red. Plutonium-239 has a specific activity of .062 curies/gram, while the corresponding figure for uranium-235 is .000002.

[180] "The attenuation of gamma rays depends on the energy of the gamma ray (generally more energetic gamma rays penetrate better, though there are some exceptions), the density of electrons (generally nearly proportional to the mass density or specific gravity) and how tightly the electrons are bound to the nuclei (much more strongly for high-Z elements). The last factor is the most important, and is why lead is used in shielding." Personal communication, Professor Jonathan Katz, Department of Physics, Washington University in St. Louis, August 7, 2008.

[181] As an analogy, when one billiard ball strikes another squarely, the first transfers its energy to the second and stops, while the second moves with about the same speed and direction as the first. By contrast, if a billiard ball strikes a bowling ball squarely, the bowling ball will move forward slightly and the billiard ball will bounce back with nearly the same velocity with which it struck the bowling ball.

[182] It takes .074 cm of lead to block half the gamma rays with an energy of 186 keV. Thus, 1 inch (2.54 cm) of lead has 2.54/.074 = 32.3 such thicknesses for 186-keV gamma rays, so (½) to the 32.3 power, or 1.9 x 10⁻¹⁰, of these gamma rays will penetrate 1 inch of lead. One kg of HEU emits 4 x 10⁷ gamma rays per second at 186 keV, ignoring absorption of the gamma rays by the uranium. (Source: Roger Byrd et al., "Nuclear Detection to Prevent or Defeat Clandestine Nuclear Attack," IEEE Sensors Journal, August 2005, p. 594.) Accordingly, one 186-keV gamma ray photon could be expected to escape 1 kg of HEU surrounded by an inch of lead every 500 seconds or so. Absorption of gamma rays by uranium would reduce this number considerably. Increasing the amount of U-235 in a bomb-usable shape would not affect this calculation much because most gamma rays would be absorbed by the uranium and the amount of lead would increase as the surface area of the uranium lump increased. Further, gamma rays radiate in all (continued...)

---

from U-235 have so little energy that many are absorbed by the uranium itself, a process known as self-shielding, so that the number of gamma rays emitted by a piece of U-235 depends on surface area, not mass.

Unclassified demonstrations performed at Los Alamos National Laboratory for the author in June 2006 indicate how shielding and self-shielding impair the detection of low-energy gamma rays from HEU. The demonstrations used a top-of-the-line detector that had an excellent ability to identify materials by their gamma-ray spectra.[183] In the first demonstration, the detector picked up gamma rays from a thin sheet of HEU foil at perhaps 30 feet away and quickly identified them as coming from HEU. The foil had a large surface area and little thickness, so there was little self-shielding. In the second, the detector gradually picked up gamma rays from a marble of HEU as it was brought closer to the detector. Because the marble had much more thickness and much less surface area than the HEU foil, there was considerable self-shielding, greatly reducing the gamma-ray output. In the third, the marble of HEU was placed in a capsule of a high-Z material, lead, perhaps 1 cm thick, and the detector picked up nothing even when the capsule was touching the window of the detector.

Sources of radiation other than SNM complicate detection. Background radiation from naturally occurring radioactive material, such as thorium, uranium, and their decay products such as radon, is present everywhere, albeit often in trace amounts. Cosmic rays generate low levels of neutrons. Some legitimate commercial goods contain radioactive material, such as ceramics (which may contain uranium), kitty litter (which may contain thorium and uranium), and gas mantles made of thorium oxide. Other radioactive isotopes are widely used in medicine and industry. Finally, a terrorist group might conceivably place radioactive material in a shipment containing a weapon or SNM chosen so as to mask the unique gamma-ray spectrum of SNM by presenting a spectrum of several known innocuous materials with peaks to interfere with those of SNM or that have an intensity much higher than SNM.

Signatures of Plutonium, Highly Enriched Uranium, and Nuclear Weapons

For purposes of this report, a signature is a property by which a substance (in particular, SNM) may be detected or identified. A nuclear weapon or its fissile material may be detected by various signatures, some of which are discussed next.

Atomic number and density

Atomic number, abbreviated "Z," is the number of protons in an atom's nucleus. For example, the Z's of beryllium, iron, and uranium are 4, 26, and 92, respectively. Z is a property of individual atoms. In contrast, density is a bulk property, and is expressed as mass per unit volume, e.g., grams per cubic centimeter. The densities of beryllium, iron, and uranium are 1.848, 7.874, and 19.050 g/cc, respectively. At its most basic, density measures how many neutrons and protons

---

(...continued)

directions. Since most detectors do not surround the object to be inspected, such as a cargo container, it would capture only a part of these gamma rays, further reducing the probability of detection.

[183] The detector used high purity germanium and was cooled with liquid nitrogen.

(which constitute almost all of an atom's mass) of a substance are packed into a volume. In general, the densest materials are those of high Z. These properties may be used to detect uranium and plutonium. Uranium is the densest and highest-Z element found in nature (other than in trace quantities); plutonium has a slightly higher Z (94), and its density varies from slightly more to slightly less than uranium, depending on its crystal structure. Some detection methods discussed in Chapter 2, such as effective Z, make use of Z; and some, such as radiography and muon tomography, make use of Z and density combined.

Opacity to photons

An object's opacity to a photon beam depends on its Z and density, the amount of material in the path of the beam, and the energy of the photons. Gamma rays and x-rays can penetrate more matter than can lower-energy photons, but dense, high-Z material absorbs or scatters them. Thus a way to detect an object, such as a bomb, in a container is to beam in x-rays or gamma rays to create a radiograph (an opacity map) like a medical x-ray.

Presence of gamma rays beyond background levels

Background gamma radiation is ubiquitous. Since many materials, including SNM, emit gamma radiation, elevated levels of gamma radiation may or may not indicate the presence of SNM.

Presence of neutrons beyond background levels

Cosmic rays and naturally occurring uranium generate a very low background flux of neutrons. Most materials do not emit neutrons spontaneously, but HEU and plutonium do. The spontaneous emission rates for 1 kg of plutonium and 1 kg of of HEU are 60,000 neutrons per second and 3 neutrons per second, respectively.[184] As a result, neutrons above the cosmic ray background coming from a cargo container would be suspicious.[185] For HEU, however, the rate is not too different from the background and thus is not a strong signature.

Gamma ray spectra

Each isotope has a unique gamma ray spectrum. For example, uranium-235 produces gamma ray peaks at several dozen discrete energy levels. This spectrum of energies is well characterized for each isotope, and is the only way to identify a particular isotope outside a well-equipped laboratory. As a result, any detector with a variant of "spectrum" or "identify" in its name, such as Advanced Spectroscopic Portal or radioactive isotope identification device, relies on identifying isotopes by their gamma-ray spectra.

---

[184] Roger Byrd et al., "Nuclear Detection to Prevent or Defeat Clandestine Nuclear Attack," *IEEE Sensors Journal,* August 2005, p. 594.

[185] "There is an important caveat in this statement, and that is cargo containers at sea. The so-called "ship-effect" that results from higher neutron levels aboard ships due to cosmic ray interactions with iron and other ship contents can result in spurious neutron readings from cargo containers at sea." Information provided by Defense Threat Reduction Agency, personal communication, August 5, 2008. Another caveat is that while innocent neutron sources other than background are rare, there are some, such as californium-252, which is produced in nuclear reactors and is used as a laboratory neutron source.

HEU presents other gamma ray signatures as well. HEU contains some U-238, which produces a gamma-ray peak at an energy of 1.001 MeV. While these gamma rays are energetic, they would be hard to detect unless the detector is very close to the uranium because they are emitted at a very low rate, and could easily be missed because trace amounts of naturally occurring uranium, such as in clay and soil, also generate 1.001 MeV gamma rays. HEU derived from spent nuclear reactor fuel rods also contains small amounts of uranium-232, which is formed when uranium is bombarded with neutrons in a nuclear reactor. Uranium-232 decays through a long decay chain of short-lived isotopes to thallium-208, which has a gamma ray of 2.614 MeV, one of the highest-energy gamma rays produced by radioactive decay, so it is distinctive as well as highly penetrating; it takes 2.041 cm of lead to attenuate half the gamma rays of that energy. Thallium-208 is also a decay product of naturally occurring thorium-232. U-232 decays very much faster than U-235 or U-238 (half-lives of 69 years, 700 million years, and 4.5 billion years, respectively), and thallium-208 decays even faster (half-life of 3 minutes), so even a very small amount of U-232 produces many gamma rays.[186] Similarly, WGPu presents various gamma ray signatures because it is a mix of several isotopes of plutonium and their decay products.

Time pattern of neutrons and gamma rays

SNM is unique in that it can fission when struck by low-energy ("thermal") neutrons. Like some other materials, it also fissions when struck by high-energy gamma rays. In a sufficiently large mass of SNM, the neutrons (usually two or three) released by the fission of one atom cause other atoms to fission, releasing more neutrons in a chain reaction.[187] SNM also fissions spontaneously, and neutrons released by these fissions have a non-negligible probability of causing other SNM atoms to fission. Characteristic products of fission offer indications that SNM is present. These products include neutrons that may be emitted over periods ranging from nanoseconds to many seconds, whether as a result of spontaneous fission or of fission induced by gamma rays or neutrons, and gamma rays emitted within nanoseconds of induced fission.

Prompt gamma rays and neutrons

When U-235 and Pu-239 fission, they release a nearly instantaneous burst of 2 or 3 neutrons and 6 to 10 gamma rays. These prompt neutrons are emitted in a continuum of energies, with an average of about 1 to 2 MeV, and are termed fast or high-energy neutrons. The prompt gamma rays are also emitted in a spectrum of many narrow lines. Only SNM will fission when struck by low-energy neutrons, so a beam of low-energy neutrons that results in a burst of neutrons and gamma rays indicates the presence of SNM. A beam of high-energy gamma rays (with energy greater than 6 MeV) will also cause SNM to fission. However, that beam will also cause other materials to fission, including natural uranium, so emission of a burst of neutrons and gamma

---

[186] For data on the properties of isotopes, see Lawrence Berkeley National Laboratory. Berkeley Laboratory Isotope Project. "Exploring the Table of Isotopes," http://ie.lbl.gov/education/isotopes htm. For further information on uranium and its decay, shielding, and detection, see footnote 8.

[187] Not every neutron will cause further fissions. For example, some may escape the mass of SNM, and some may strike impurities. In a subcritical chain reaction, each neutron results in fissions that generate, on average, less than one additional neutron; this reaction dies out. In a critical chain reaction, each neutron results in fissions that, on average, produce one additional neutron. In a supercritical chain reaction, each neutron results in fissions that, on average, produce more than one additional neutron. Such a reaction may be controlled, as in a nuclear reactor, releasing energy over months or years, or uncontrolled, as in an atomic bomb, releasing vast amounts of energy in a fraction of a second.

---

rays resulting from interrogation by a high-energy gamma ray beam is a possible, but not a definitive, indicator by itself of the presence of SNM.

Delayed gamma rays and neutrons

When U-235 or Pu-239 atoms fission, they split into two smaller fission fragments in any of approximately 40 ways for each isotope, resulting in "[s]omething like 80 different fission fragments" for U-235 or Pu-239.[188] These fission fragments are unstable and decay radioactively into isotopes of various elements. Fission is a statistical process, so that fissioning of a great many U-235 or Pu-239 atoms produces a complex mixture of some 300 isotopes of 36 elements.[189] These isotopes have a great range of half-lives, from a small fraction of a second to millions of years, but the isotopes with a half-life greater than approximately 30 years emit only very low levels of radiation. This process produces thousands of times more gamma rays than neutrons. Since much cargo consists of low-Z material and since gamma rays penetrate low-Z cargo much more readily than do neutrons, many more gamma rays than neutrons resulting from fission of SNM escape containers holding such cargo. Higher-Z cargo will attenuate the gamma rays more than the neutrons. Some of the gamma rays have energies exceeding those of thallium-208, 2.614 MeV, the highest energy typically observed in natural backgrounds. "Their high energy makes this gamma radiation a characteristic of fission, very distinct from normal radioactive background that typically produces no gamma radiation exceeding an energy of 2.6 MeV."[190] Note that some other isotopes, such as U-238 and Pu-240, are "fissionable," that is, they can undergo fission only when struck by high-energy (fast) neutrons. The high-energy gamma rays resulting from fission are a strong indicator of the presence of SNM.[191] The intensity of the neutron and gamma-ray flux over a short period, caused by rapid decay of many of the fission products, and the prompt response to a probe, are distinctive signatures as well.

There is another time-delay signature. A neutron beam makes atoms of some other elements radioactive, in particular transforming some atoms of stable oxygen-16 to radioactive nitrogen-16. Researchers at Lawrence Livermore National Laboratory conducted experiments in which they bombarded a target of natural uranium (99.3% U-238, 0.7% U-235) inside a cargo container with a neutron beam, and recorded the gamma ray spectrum resulting from radioactive decay. After they turned off the neutron beam, they found that the high-energy portion of the spectrum was dominated by gamma rays from the decay of nitrogen-16 for the first 15 seconds, and after that the dominant signal was from the decay of radioactive fission products, with an average half-life of about 55 seconds.[192] This time difference is an indicator of the presence of SNM.

---

[188] U.S. Department of Defense and Department of Energy. *The Effects of Nuclear Weapons,* Third Edition, compiled and edited by Samuel Glasstone and Philip Dolan, Washington, U.S. Govt. Print. Off., 1977, p. 633.

[189] Ibid.

[190] D.R. Slaughter et al., "The 'Nuclear Car Wash': A Scanner to Detect Illicit Special Nuclear Material in Cargo Containers," UCRL-JRNL-202106, January 30, 2004, p. 4.

[191] They are not a definitive indicator, however, because there could be other sources of fission, such as californium-252, and cosmic rays could induce fission.

[192] D.R. Slaughter et al., "The 'Nuclear Car Wash': A Scanner to Detect Illicit Special Nuclear Material in Cargo Containers," UCRL-JRNL-202106, January 30, 2004, pp. 6-7.

Differential die-away

Interrogation of SNM with a beam of neutrons or high-energy photons to induce fission produces another unique signature. While the beam may cause neutrons to be emitted immediately through various nuclear reactions (e.g., fission), materials other than SNM will not support a nuclear chain reaction. In contrast, even a subcritical mass of SNM can sustain a chain reaction for a short time. As a result, fission in a multi-kilogram block of SNM will continue to produce neutrons for a short time after the beam has been turned off, with the intensity and duration of the neutron flux depending on the amount of SNM and the cargo loading. This delayed neutron time signature is called differential die-away, is measured on the order of a thousandth of a second after the beam is turned off, and is specific to U-235 and Pu-239 (and, rarely, other fissile isotopes). This technique depends on the detection of the prompt fission signal, but hydrogenous materials such as those found in cargo tend to attenuate this signal, and there may be background neutrons, so that some difficult scans may require more time, possibly two minutes, and some may not be feasible.

Fission chain time signature

A subcritical mass of SNM is too small to support a supercritical chain reaction because too many neutrons escape the SNM for the number of neutrons to increase exponentially. Nonetheless, chain reactions do occur in SNM, triggered by a neutron from spontaneous fission or a background neutron. These chain reactions may last several to dozens of generations, producing a burst of neutrons and gamma rays over some billionths of a second. No other material produces this signature. In contrast, most background neutrons and gamma rays arrive at a detector in a random pattern. The one exception is that neutrons generated as cosmic rays strike matter also tend to be generated in bursts; work is under way to try to differentiate between bursts of neutrons induced by cosmic rays and those generated by fission chains. Detection of this signature is therefore a strong sign of the presence of SNM. Unlike differential die-away or delayed neutrons and gamma rays, this signature can be detected with passive means provided the SNM is not well shielded. This technique places great demands on detector technology but can be done with state-of-the-art electronics.

Chapter 2 discusses in detail two other signatures—deflection of muons and nuclear resonance fluorescence and absorption—and their detection.

Detecting Signatures of a Nuclear Weapon or SNM

Overview: How are signatures gathered, processed, and used?

Detection involves using detector elements to gain data, converting data to usable information through algorithms, and acting on that information through concept of operations, or CONOPS. Detectors, algorithms, and CONOPS are the eyes and ears, brains, and hands of nuclear detection: effective detection requires all three.

Since photons or neutrons have no electrical charge, their energy is converted to electrical pulses that can be measured. This is the task of detectors, discussed next. The pulses are fed to algorithms. An algorithm, such as a computer program, is a finite set of logical steps for solving a problem. For nuclear detection, an algorithm must process data into usable information fast enough to be of use to an operator. It receives data from a detector's hardware, such as pulses representing the time and energy of each photon arriving at the detector. It converts the pulses to a format that a user can understand, such as displaying a gamma ray spectrum, identifying the material creating the spectrum, or flashing "alarm." Every detector uses one or more algorithms. Improvements to algorithms can contribute as much as hardware improvements to detector capability. Algorithms may be improved in many ways, such as by a better understanding of the physics of a problem, or by improving the computers in detection equipment so they can process more elaborate algorithms.

CONOPS may be divided into two parts. One specifies how a detection unit is to be operated to gain data. How many containers must the unit scan per hour? How close would a detector be to a container? Shall the unit screen cargo in a single pass, or shall it be used for primary screening, with suspicious cargo sent for a more detailed secondary screening? A second part details how the data are to be used. What happens if the equipment detects a possible threat? Which alarms are to be resolved on-site and which are to be referred to off-site experts? Under what circumstances would a port or border crossing be closed? More generally, how is the flow of data managed, in both directions?[193] What types of intelligence information do CBP agents receive, and how do data from detection systems flow to federal, state, and local officials for analysis or action? While this report does not focus on CONOPS because is not a technology, it is an essential part of nuclear detection.

How detectors work

A discussion of how detectors work is essential to understanding the capabilities and limits of current detectors and how detectors may be improved. Detecting each signature of a nuclear weapon or SNM requires a detector that is appropriate for that signature. Further, there is a hierarchy of gamma ray detectors. The simplest can only detect the presence of gamma radiation. The next step up, detectors with low energy resolution, have a modest capability to identify an isotope by its gamma ray spectrum. Next, detectors with high energy resolution have very accurate isotope identification capabilities. More sophisticated detector systems can also identify the presence of SNM by the time pattern of gamma rays released when such material fissions. The most sophisticated detector systems can produce an image showing where each gamma ray came from.

Detectors require a signal-to-noise ratio sufficient to permit detection. That is, they must be able to extract the true signal (such as a gamma-ray spectrum) from noise (spurious signals caused, for example, by background radioactive material or by imperfect detectors or data-processing algorithms). Two concepts are central to gamma-ray detector sensitivity: detection efficiency and spectral resolution. The former refers to the amount of signal a detector records. One aspect of detection efficiency is the fraction of the total emitted radiation that the detector receives. Radiation diminishes according to an inverse square law; that is, the intensity of radiation (e.g., number of photons per unit of area) from a source is inversely proportional to the square of the distance from the source.[194] Since a lump of SNM emits radiation in all directions, moving a detector closer to SNM, or increasing its size, increases efficiency. Reducing the cost of the active material in a detector may increase efficiency by making a larger sensor area affordable. Another aspect is the fraction of the radiation striking the detector that creates a detectable signal. For

---

[193] For further analysis of this topic, see CRS Report RL34070, *Fusion Centers: Issues and Options for Congress*, by John Rollins.

[194] See, for example, "Inverse Square Law, General," at http://hyperphysics.phy-astr.gsu.edu/Hbase/Forces/isq.html.

example, a detector that can absorb 90% of the energy of photons striking it is more efficient than one that can absorb 10%. A more efficient detector will collect information faster, reducing the time it takes to screen a cargo container.

Spectral resolution refers to the sharpness with which a detector presents energy peaks in a radiation spectrum. A graph of the gamma-ray spectrum of a radioactive isotope plots energy levels along the horizontal axis of the graph and the number of counts per unit time at each energy level along the vertical axis. A perfect device would record the energy levels of a gamma-ray spectrum as a graph with vertical "needles" of zero width because each radioactive isotope releases gamma rays only at specific energy levels. In practice, however, detectors are not perfect, and 186-keV gamma rays will be recorded as a bell curve centered on 186 keV. The narrower the spread of the bell curve,[195] the more useful the information is. Polyvinyl toluene (PVT), a plastic that is widely used in radiation detectors because it can be fielded in large sheets at low cost, is sensitive but has poor resolution, i.e., extremely wide bell curves for each gamma-ray energy level. As a result, while PVT can detect radiation, the peaks from gamma rays of different energy levels blur together, making it impossible to identify an isotope. **Figure 3** makes this point; it shows the spectra of 90% U-235 and background radiation as recorded by a PVT detector. At the other extreme, high-purity germanium (HPGe) produces very sharp peaks, permitting clear identification of specific isotopes. These detectors are expensive, heavy, have a small detector area, and must be cooled to extremely low temperatures with liquid nitrogen or a mechanical system, making them less than ideal for use in the field. However, mechanically cooled HPGe detectors weighing some 2.5 kg are being developed for field use.[196] **Figure 4** shows the spectrum of Pu-239 as recorded by various types of detectors with better resolution than PVT.

Various means can improve detector sensitivity.[197] One type of semiconductor detector crystal is cadmium-zinc-telluride, or CZT. The peak on the far right of each spectrum[198] in **Figure 5** shows improvement in the resolution of the gamma-ray spectrum for cesium-137 (a radioactive isotope) taken with different CZT detectors that, for the years indicated, were at the high end of sensitivity. The top line shows a spectrum taken with a CAPture device developed by eV Products (1995-1998); the middle line shows a spectrum taken with a coplanar-grid device developed by Lawrence Berkeley National Laboratory (2000-2003); and the bottom line shows a spectrum taken with a 3-D device developed by the University of Michigan (2008). Better CZT crystals and better ways to overcome limitations of these crystals have both improved sensitivity in various ways:

- Researchers have been able to grow larger crystals. CZT crystal volume for the three devices was 1.00 cc in 1995-1998, 2.25 cc in 2000-2003, and 6.00 cc at present. Larger crystals are more efficient, i.e., they can capture more photons,

---

[195] The spread is measured as "full width at half maximum," that is, the width of the curve measured halfway from the top to the bottom of the curve.

[196] Personal communication, Defense Threat Reduction Agency, August 8, 2008.

[197] This paragraph was prepared with the assistance of Aleksey Bolotnikov, Physicist, Brookhaven National Laboratory, Professor Zhong He, Department of Nuclear Engineering and Radiological Sciences, University of Michigan, and Ralph James, Senior Physicist, Brookhaven National Laboratory, July 2008.

[198] The peak in the energy spectrum corresponds to the total energy of an incoming photon completely stopped inside the detector. The continuum on the left side of the peak is caused when an incoming photon scatters in the detector, depositing an unpredictable fraction of its total energy. For such events, the information about the photon's energy is lost. Such events contribute to the background that may affect detector performance.

and more of the energy of individual photons, permitting more counts per unit time (i.e., more data).

- Crystal quality has improved. A more uniform crystal structure and fewer impurities allow for better transport of the photon-induced electrical charge through the crystal and thus more accurate determination of the energy of each photon.

- University of Michigan researchers have constructed three-dimensional arrays of CZT crystals, permitting their detector to determine the 3-D coordinates of each individual gamma ray photon as it interacts with the CZT crystal, in turn permitting location as well as identification of gamma-ray sources.

- Electronics have improved. Researchers have made significant progress in reducing the noise inherent in electronic circuits (application-specific integrated circuits) that translate signals from the interaction of photons with CZT into a form in which algorithms can process them. Reducing the noise in these circuits permits more accurate measurement of gamma-ray energy. For example, a circuit developed in 2007 by Brookhaven National Laboratory has improved energy resolution substantially, and other advances in detector electronics in the last few years enable electronic components to compensate for defects in the crystals (analogous to adaptive optics in astronomy).

- Algorithms to reconstruct the signal from gamma rays have improved, also permitting more accurate measurement of gamma ray energy.

Another factor that affects the ability to detect SNM is the time available for a detector to scan a container or other object, often called "integration time." Detectors build up radiography or tomography images, or gamma-ray spectra, over time. More time enables a detector to have more photons per pixel (in the case of radiography) or per bin (in the case of gamma-ray spectra), or more muons per voxel. More time also enables a neutron detector to detect more neutrons and measure their times of arrival, as discussed below, helping to determine if the neutrons are generated by SNM or by background materials. More time thus provides better data, which provides for better separation of signal from noise, better separation of different sources of radiation, fewer false alarms, and a better chance of detecting and identifying shielded threat material. **Figure 16** illustrates how one system builds up an image over time. From a physics perspective, then, increasing integration time improves the accuracy of the result, but from a port operator's perspective, longer integration time impedes the flow of commerce, which costs money, so a balance must be struck between these two opposed goals. This balance may be stated formally in a concept of operations (discussed in more detail below), which specifies how, among other things, a detection system will be operated; detection equipment must be designed to operate within the time required, and port operations must allow that amount of time for scans.

Still another means of improving the ability to detect SNM is to increase the spatial resolution of a detector. According to DTRA,

> This is easily demonstrated in the example of a shielded versus unshielded radiation detector. Unshielded detectors are sensitive to radiation impinging on it in all directions, which is characteristic of the nature of naturally-occurring background radiation. By adding shielding,

a detector's field-of-view can be controlled, and background radiation levels reduced, increasing the signal-to-noise ratio for the detector in the direction the detector is aimed.[199]

Detecting gamma rays

Gamma rays do not have an electrical charge, but an electrical signal is needed to measure them. There are two main ways by which a gamma ray can be turned into electrical energy. One is with a scintillator material, such as PVT or sodium iodide. When a single higher energy photon, such as a gamma ray, strikes the scintillator and interacts with it, the scintillator emits a large number of photons of lower energy, usually visible light ("optical photons"). A photomultiplier tube (PMT) converts the optical photons to electrons, then multiplies the electrons to generate a measurable pulse of electricity whose voltage is proportional to the number of optical photons, which is in turn proportional to the energy deposited by the gamma ray. An electronic device called a multi-channel analyzer sorts the pulse into a "bin" depending on its energy and increases the number of counts in that bin by one. A software package then draws a histogram with energy level on the horizontal axis and number of counts on the vertical axis. The histogram is the gamma ray spectrum for that isotope.

In contrast, a semiconductor material, such as HPGe, turns gamma rays directly into an electrical signal proportional to the gamma-ray energy deposited. A voltage is applied across the semiconductor material, with one side of the material being the positive electrode and the other being the negative electrode. When a gamma ray strikes the material, it knocks some electrons loose from the semiconductor's crystal lattice. The voltage sweeps these electrons to the positive electrode. Their motion produces an electric current whose voltage is proportional to the energy of each gamma ray. Each pulse of current is then sorted into a bin depending on its voltage and the spectrum is computed as described above.[200]

This approach, with either type of detector, is used to detect the various gamma-ray signatures described earlier. However, the requirements for detecting time signatures varies somewhat. Because prompt gamma rays are emitted so quickly, identifying them requires the ability to record time of arrival to several billionths of a second. Delayed gamma rays of interest are generated over a period of tens of seconds, so the ability to record precise time of arrival is less important. Detecting fission chain time signature requires a high-efficiency detector because long fission chains are relatively rare. Thus to detect SNM rapidly, the detector must have a high efficiency for detecting every fission chain. While the delayed emissions from fission chains are too weak to detect passively, fission chain time signature focuses on detecting the prompt emissions from fission, which are stronger. High efficiency is also important for neutron and gamma-ray interrogation, but the emphasis is less stringent because far more fissions are induced (i.e., the signal is stronger). Detecting nuclear resonance fluorescence requires high-resolution detectors in order to differentiate between the various materials being analyzed.

---

[199]

Personal communication, August 8, 2008.

[200] For a simple discussion of how semiconductors work, see Marshall Brain, ""How Semiconductors Work," at http://www.howstuffworks.com/diode htm. For further detail, see Knoll, *Radiation Detection and Measurement*, Chapter 11, "Semiconductor Diode Detectors," pp. 353-403.

Detecting neutrons

Neutrons, like photons, do not have an electrical charge, but the two interact with matter differently. Photons interact chiefly with electrons, while neutrons interact with atomic nuclei. As a result, neutrons are counted by a different process. A common neutron detector is a tube of helium-3 gas connected to a high-voltage power supply, with positively and negatively charged plates or wires in the tube. In its rest state, current cannot pass through the helium because it acts as an insulator. When a low-energy neutron passes through the tube, it is absorbed by a helium-3 atom, producing a triton (1 proton and 2 neutrons) and a proton. These particles are highly energetic and lose their energy by knocking electrons off other helium-3 atoms. Positively charged ions of helium-3 move to the negative plate, while electrons move to the positive plate. Since electric current is the movement of charged particles, these particles generate a tiny electric current that is measured and counted. Neutrons are emitted as a continuum of energies. While the mean energy of each neutron spectrum varies somewhat from one isotope to the next, neutron energy spectra do not have lines representing discrete energies as with gamma rays. Moreover, neutrons lose energy as they collide with low-Z material, further blurring their spectra. Thus neutron spectra are of little value for identifying isotopes. Instead, the total neutron count is an important means of identifying SNM because only SNM gives off neutrons spontaneously in significant numbers, though some neutron background is generated mainly when cosmic rays knock neutrons off atoms. Several other methods of detecting SNM by neutron emission, discussed above, rely on the time pattern in which a group of neutrons arrives.

Several systems detect neutrons with tubes filled with helium-3 (He-3), a standard method. DOE obtains He-3 as a byproduct from the decay of tritium used in nuclear warheads. With the decades-long decline in numbers of warheads and a hiatus in tritium production for many years, there is little new supply of He-3. DOE plans to supply customers with 10,000 liters of He-3 a year, with a starting bid price expected to be around $72 per liter, and states, "This appears short of what customers are requesting."[201] (Russia sells He-3 to U.S. companies, but quantities are proprietary and not available.) Deploying He-3 neutron detection systems in large numbers would require a considerable amount of He-3. Customs and Border Protection (CBP) states that "based on our RPM [radiation portal monitor] deployments CBP would need approximately 2500 [detector] units to cover sea and land borders."[202] (Data for number of units needed for air cargo are not available.) Given the shortage and cost of He-3, deployment of neutron detectors using large amounts of He-3, or large numbers of units requiring small amounts of He-3, does not appear feasible.[203]

Alternative neutron detection systems are possible. They include tubes coated with boron-10 or lithium-6, tubes filled with boron-10 trifluoride (a toxic gas), nanocomposite scintillators, and "neutron straws," thin tubes being developed commercially under sponsorship of the Defense Threat Reduction Agency.[204] Substituting any of these technologies for He-3 in a system would necessitate re-engineering the system's neutron detectors, revising algorithms, conducting tests, perhaps modifying the resulting system for operational conditions, and so on. Those changes have

---

[201] Information provided by Isotope Program, U.S. Department of Energy, personal communication, June 30, 2008.

[202] Information provided by Customs and Border Protection, Department of Homeland Security, personal communication, July 25, 2008.

[203] The American Association for the Advancement of Science held a workshop on the helium-3 shortage on April 6, 2010. Briefing slides are available at http://cstsp.aaas.org/agenda_meeting.html.

[204] See Proportional Technologies, Inc., "Neutron Straws," at http://www.proportionaltech.com/neutron.htm.

the potential to add delay and affect system performance (for better or worse), though given the high cost of He-3 (about $2.7 million for 38,000 liters) they might well reduce cost.

Detecting absorption or scattering of high-energy photons

Photons of sufficiently high energy can penetrate solid objects. Denser, higher-Z material within a solid object absorbs photons of lower energy and scatters photons of higher energy. For cargo scanning, a fan-shaped planar beam of photons is sent through a cargo container as the container passes through the beam, and a detector array on the other side consisting of semiconductor or scintillator material records the opacity of each pixel to the photons. An algorithm then creates a two-dimensional opacity map of the contents of the container and displays it as an image on a computer screen.

Increasing the energy of photons allows them to penetrate more material. Radiography is used to search cargo containers for terrorist nuclear weapons, among other things.[205] A radiograph would reveal clearly a large dense object, such as a nuclear weapon encased in lead shielding. Two limitations of radiography are noteworthy. First, radiographs do not detect radiation and thus do not specifically detect SNM, just high-density, high-Z material. Second, if a terrorist bomb is placed in a shipment of dense or mixed objects, the image of the bomb might be hidden or a radiographic equipment operator might not notice it. It would be much harder to detect a small piece of SNM using radiography than to detect a bomb.

Evasion of Detection Technologies

In order to understand the capabilities of detection systems, it is important to know their weaknesses as well as their strengths. However, detailed discussions of means of evasion tend to become classified. Some references are made throughout this report, but some are withheld to keep the report unclassified. In general, an enemy could use various means in an effort to defeat these technologies. For example, high-Z material absorbs and deflects gamma rays, low-Z material deflects neutrons, radiography might miss a small piece of SNM (especially if mixed in with other dense material), and reducing the apparent density and Z of SNM by mixing it with a low-Z substance reduces the deflection of muons.

Further, enemy attempts to defeat one type of detection system may complicate plans or make a plot more vulnerable to detection by other means, as several examples illustrate. (1) The use of multiple detection systems that detect different phenomena are harder to defeat than those detecting one phenomenon only. Placing a lead shield around a bomb in order to attenuate gamma rays from plutonium would create a large, opaque image that would be quite obvious on a radiograph. It is for this reason that Congress mandated, "A container that was loaded on a vessel in a foreign port shall not enter the United States (either directly or via a foreign port) unless the container was scanned by nonintrusive imaging equipment and radiation detection equipment at a foreign port before it was loaded on a vessel." This restriction is to apply by July 1, 2012.[206] (2) An enemy could attempt salvage fuzing, which would detonate a weapon if the weapon sensed attempts to detect it, such as with photon beams, or if it was tampered with. However, salvage

---

[205] See, for example, Katz, J. I., Blanpeid, G. S., Borozdin, K. N. and Morris, C., "X-Radiography of Cargo Containers," *Science and Global Security,* Vol. 15, 1: 49-56.

[206] P.L. 110-53, Implementing Recommendations of the 9/11 Commission Act of 2007, Section 1701, 121 Stat. 489.

fuzing has various shortcomings. It could result in a weapon detonating by accident or if it is scanned overseas. It could detonate a weapon in a U.S. port, where it would do much less damage than in a city center. (3) Enemy attempts to smuggle HEU into the United States in order to avoid detection of a complete bomb would require fabricating the weapon inside this nation, which in turn could require such activities as smuggling other weapon components and purchase of specialized equipment, and could run the risk of accidents (such as with explosives), any of which could provide clues to law enforcement personnel. For these reasons, it is important to view technology development not only as advances in capabilities per se but also in the context of an offense-defense competition.

Author Contact Information

Jonathan Medalia
Specialist in Nuclear Weapons Policy
jmedalia@crs.loc.gov, 7-7632

www.ingramcontent.com/pod-product-compliance
Lightning Source LLC
Chambersburg PA
CBHW081502170526
45166CB00008B/2523